ROUTLEDGE LIBRA
ECONOMIC GE(

T0227442

Volume 2

TOWARDS GLOBAL LOCALIZATION

TOWARDS GLOBAL LOCALIZATION
The computing and telecommunications industries in Britain and France

PHILIP COOKE, FRANK MOULAERT,
ERIK SWYNGEDOUW, OLIVIER WEINSTEIN
AND PETER WELLS

WITH MARTINE LEMATTRE AND
PATRICE GREVET

Routledge
Taylor & Francis Group

LONDON AND NEW YORK

First published in 1992

This edition first published in 2015
by Routledge
2 Park Square, Milton Park, Abingdon, Oxon, OX14 4RN

and by Routledge
711 Third Avenue, New York, NY 10017

Routledge is an imprint of the Taylor & Francis Group, an informa business

British Library Cataloguing in Publication Data
A catalogue record for this book is available from the British Library

ISBN: 978-1-138-85764-3 (Set)
eISBN: 978-1-315-71580-3 (Set)
ISBN: 978-1-138-85783-4 (Volume 2)
eISBN: 978-1-315-71839-2 (Volume 2)
Pb ISBN: 978-1-138-85784-1 (Volume 2)

Publisher's Note
The publisher has gone to great lengths to ensure the quality of this reprint but points out that some imperfections in the original copies may be apparent.

Disclaimer
The publisher has made every effort to trace copyright holders and would welcome correspondence from those they have been unable to trace.

TOWARDS GLOBAL LOCALIZATION

The computing and
telecommunications industries
in Britain and France

Philip Cooke
Frank Moulaert
Erik Swyngedouw
Olivier Weinstein
Peter Wells

with
Martine Lemattre
Patrice Grevet

UCL
PRESS

© Philip Cooke, Frank Moulaert, Erik Swyngedouw,
Olivier Weinstein, Peter Wells,
Martine Lemattre, Patrice Grevet 1992

First published in 1992 by UCL Press

UCL Press Limited
University College London
Gower Street
London WC1E 6BT

The name of University College London (UCL) is a registered
trade mark used by UCL Press with the consent of the owner.

ISBN: 1-85728-000-8

A CIP catalogue record for this book is available from the British Library.

This book was typeset using word-processing software exclusively
and was produced as camera-ready copy on a laser printer.

First published in paperback
in 1993 by UCL Press
ISBN: 1-85728-261-2
(Hardback ISBN: 1-85728-000-8, as above)

Typeset in Times Roman and Univers.
Printed by Biddles Ltd, King's Lynn and Guildford, England.

CONTENTS

		Page
	Preface	vi
1	Computing and communications in the UK and France: innovation, regulation and spatial dynamics – an introduction *Philip Cooke*	1
2	High technology and flexibility *Olivier Weinstein*	19
3	Accumulation and organization in computing and communications industries: a regulationist approach *Frank Moulaert & Erik Swyngedouw*	39
4	Globalization and its management in computing and communications *Philip Cooke & Peter Wells*	61
5	The regional patterns of computing and communications industries in the UK and France *Erik Swyngedouw, Martine Lemattre, Peter Wells*	79
6	The computer hardware industry in the 1980s: technological change, competition and structural change *Peter Wells & Philip Cooke*	129
7	The telecommunications equipment industry: the great transformation *Olivier Weinstein*	152
8	Services: the bridge between computing and communications *Frank Moulaert*	178
9	Global localization in computing and communications: conclusions *Philip Cooke*	200
	References	215
	Index	225

vii

CONTENTS

Page

Preface

1   Computing and communications in the UK and France:
     innovation, regulation and spatial dynamics,
     an introduction   *Felix Clark*   1

2   High technology and flexibility *Oliver Weinstein*   45

3   Accumulation and organization in computing and
     telecommunications industries: a regulationist approach
     Frank Moulaert & Eric Swyngedouw   59

4   Globalization and its management in computing and
     communications *Philip Cooke & Peter Wells*   81

5   The regional pattern of computing and communications
     industries in the UK and France
     Erik Swyngedouw, Martine Lazennky, Peter Wells   99

6   The computer hardware industry in the 1980s:
     technological change, competition and
     regional change
     Peter Wells & Philip Cooke   120

7   The telecoms/multicore equipment industry:
     the great transformation *Oliver Weinstein*   152

8   Services: the bridge between computing
     and communications *Pascal Monnoyer*   175

9   Global localization in computing and
     communications: conclusions *Philip Cooke*   200

     References   215
     Index

For Jack Dyckman

PREFACE

This book is dedicated to Jack Dyckman who died in the spring of 1987. Jack's last post had been Director of the Johns Hopkins European Centre for Regional Planning and Research at Lille in France. The research project on which this book is based had its origins in a joint venture between the Centre and the Department of City and Regional Planning at the University of Wales College of Cardiff. The occasion of the project's birth was an evening at the Brasserie de la Paix, hosted with typical generosity by Jack at his favourite restaurant in Lille.

Jack's death occurred between the submission of the application for funding to the Franco-British scheme of the CNRS and ESRC and the announcement of the successful outcome of the bid. We are grateful to both research councils for their support. We also hope Jack would have been pleased with the final product of his and our efforts.

The research itself was led from Britain by Philip Cooke and from France by Frank Moulaert, the Deputy-Director of the Johns Hopkins Centre, aided by Erik Swyngedouw who held an appointment there. Subsequently, Olivier Weinstein, Martine Lemattre and Patrice Grevet from the University of Lille joined the French team. Peter Wells became the research fellow on the British side. The cross-national team proved to be an exceptionally cordial group of colleagues and the experience of doing research to a tightly defined common format in two strikingly different countries was a rewarding and, in many respects, a formative experience. Many of us have subsequently engaged in other cross-national research projects, enthused by the stimulus of international collaboration.

Many colleagues helped us to develop our thinking and to deal with technical or theoretical problems on the way to completing the research. We would like to thank especially Kevin Morgan, Alain Lipietz, Richard Florida, Gordon Clark, Benjamin Coriat, Louis Albrechts, Luc Soete, Ash Amin, Peter Van Hoogstraten, Anna Lee Saxenian and François Bar, who at various times commented on aspects of our work. Above all, we wish to thank the business managers of firms we interviewed who, without exception, gave us much more of their time than we expected to tell us what was going on in the fast-moving industries we had chosen to study. We would like to think that all who knew him would be happy that the result of our collective efforts is dedicated to Jack Dyckman.

Philip Cooke, Frank Moulaert, Erik Swyngedouw
Olivier Weinstein, Peter Wells

CHAPTER ONE
Computing and communications in the UK and France: innovation, regulation and spatial dynamics – an introduction

Philip Cooke

The aims of the book

This book is a study of changes taking place in two key advanced-technology industries: computing and communications (C&C). The study is broadened by its focus on the transformation of the industries in two major European economies: Britain and France. It could be said that the two master technologies of C&C have been tending to converge in the past decade, but the policy environments of Britain and France have diverged radically during the same period. In France there has been a continuation of *dirigisme* in the fields reported here, industrially and territorially. In Britain, the 1980s saw a radical experiment in withdrawal or diminution of government support for both industries (but particularly – through privatization – of the telecommunications sector) and regions.

Thus, a major theme of this book is to what extent the dynamics of economic development and industrial reorganization are dictated by pure market forces, as against the force of government policies seeking to intervene in and shape the competitive market. The development of C&C in these two countries offers a valuable laboratory in pursuit of this aim, precisely because of the differences in the posture of state policy amid universal technological trends.

A second major aim of the book is to explore the relevance of a broad macroeconomic perspective, the "regulationist approach" (Lipietz 1987, Boyer 1988). This school of thought purports to explain the intimate relationships between particular forms of industrial organization and institutional framework that persist over time, and why they change. It is a macro-perspective that subsumes some theoretical propositions, among which those addressing questions of production flexibility, technological innovation, inter-firm institutional structures, globalization and deregulation of markets by governments are a few of the more prominent. Rather unusually, this book examines the validity of the regulationist approach by looking at both macro- and microeconomic phenomena, attempting in the most testing way to establish the empirical credentials of the perspective. Thus, the extent to which firm behaviour in either of our target sectors can be understood in terms of "regulation school" propositions is taken as at least a provisional judgement of the theoretical validity of the approach.

The third main aim of the book is to see, again with the aid of the regulationist perspective, whether a good explanation can be provided for the spatial shifts – both global and local – described in the main empirical sections of this study. It is fair to say that one of the general attractions of the regulationist approach lies in its sensitivity to spatial matters. Accordingly, the perspective has been taken up with some gusto in the spatial analytic disciplines: economic geography, regional science, urban and regional studies (see, for example, Scott & Storper 1986, Leborgne & Lipietz 1988, Moulaert et al. 1988,). Since one of the more interesting features of contemporary economic change is an apparent revaluation of the importance of geographical location and linkage, the book seeks to explain this by teasing out the interactions between organizational and technological innovation by firms, and institutional adjustments in the environmental context in which firms operate. Moreover, since policy-makers from the local to the supranational level express an acute interest in the development of these key advanced technologies, seeking to gain advantage for their territories in the competitive struggle, we hope to shed light on the processes of change, to enable appropriate policy conclusions to be drawn.

There are five key issues which form the broad structure of what is discussed in the book. First, it is necessary to draw attention to certain apparently pervasive technological changes which have in turn led to innovations in both the products and the processes of their production. Secondly, it is vital that the institutional structures that permeate the more directly economic pursuits of firms are presented along with key changes within these pursuits. Next, attention is drawn to two important geographical scales by which it is relevant to consider the effects of these technical and institutional changes. Hence, the third issue concerns the macro-spatial level and the question of "globalization", while the fourth relates to the micro-spatial, or regional and local, shifts occasioned by the developments occurring in C&C industries. Finally, analysis of the sectoral dynamics of three crucial activities must be foreshadowed, these being the manufacture of computing and communications equipment, and the delivery of specialist value-added services to which they give rise. In what follows, each of these five core areas of the theoretical and empirical research will be outlined, and key elements of each will be highlighted.

Innovation, technical change and the firm

New products and new methods of producing them have always been crucial competitive weapons to firms. However, in the past two decades there has been a huge leap in the rate of product and process innovation, particularly in the computing and communications industries. An exponential growth of new products in telecommunications is demonstrated in Ungerer & Costello (1988), and the profile for computing is comparable, while that in communications services is sharper still and over a much shorter period (Mansell 1989).

A striking feature of innovation in these industries is the noticeable shift in the processes underpinning it. Hitherto, the predominant source of innovation has been the large hierarchical corporation with a divisional structure, one of the key departments of which is research & development (R&D). Whatever organizational changes may have been instituted within large C&C corporations in recent years – and there have been many, as they have sought to cope with the twin uncertainties of heightened competition and more rapid rates of technological change – the R&D department has survived more or less unscathed. However, its capacity for "stand-alone" innovation may have diminished. One reason for this is that innovation has become a far more interactive process than it was as a giant corporation such as IBM grew to prominence. Recent studies show a pervasive increase in the amount of corporate innovative activity occurring through interaction between the innovator and user (Hakansson 1989). This growth of "user-power" is, in itself, something of an innovation and one that has been occasioned by the greater choice a user now has in a more competitive market, and the deepening of markets as more and more users see the potential competitive advantage of investing in C&C. Hakansson (1989) shows that this interaction between client and producer is the most

3

widespread form of inter-firm co-operation in the innovation sphere, accounting for 75% of such co-operative relationships. Moreover, they are the longest-lived form of co-operation, averaging ten years durability. Among the reasons for increased user–producer collaboration in innovation are: the specificity of user needs demanding "bespoke" applications; the prospect of market enhancement through customer "lock-in", and through learning how to develop new applications for sale elsewhere; and the continuous learning that may be afforded from the user's growing ability to demand "aftercare" from the producer, once installation has been completed.

A different kind of innovative learning by interaction comes from heightened collaborative efforts between producers (Lundvall 1988). Specifically, the interaction here is likely to be between a larger customer firm and a smaller supplier firm in a similar or related component or service-supply industry. In parallel with user-power is what can be termed "supplier-power" (Cooke & Morgan 1990): this is a situation where, for reasons of increased competition, a firm is forced to reduce, say, materials costs by subcontracting production to one or more independent suppliers. Where suppliers have "strategic competences" (Teece 1986), they may be in a position to demand not the short-term multi-supplier contract, that most customers require to maintain control over the relationship, but long-term single-supplier status. To the extent that the latter position is approached, if not achieved, the relationship between customer and supplier will necessarily have changed from a competitive market relationship to one of increasing trust and confidentiality (Granovetter 1985). Such relationships can result in innovation through learning by interacting. Firms jointly develop and produce, and added value stems from the exchange of the know-how each partner possesses.

The third form of innovative learning by interacting is the inter-firm agreement between equals or near-equals, as distinct from the previous two categories. Though this is one of the most widely discussed innovation processes in the relevant literature, it is much less important than user–producer interaction and not necessarily more important than customer–supplier interaction. Nevertheless, it has attracted attention because it seems to be a new form of relationship between potentially competitive rivals (Chesnais 1988, Mytelka 1990). Strategic alliances, as such collaborations are often referred to, are a prime focus of the micro-analysis presented in this book. They are important, especially so in C&C, as forms of pre-competitive as well as competitive innovation and related activity, but also because they often presage acquisition or merger strategies which are of importance to industry dynamics and competitive position. Ultimately, as is argued in Chapter 2 of this book, innovation is a key element of "knowledge-intensive production", and, once the inter-firm interaction chains are developed, this progresses into a form of competition in which economies of permanent innovation begin to predominate.

Technological change of this permanent, in-built or embedded kind is what characterizes the innovation strategy of the leading Japanese firms, but it is not confined to them. Some leading Western corporations have, through either emulation or collaboration, learned from their interactions with the key competitor firms. The most important value-increment accrued from the quest for permanent innovation is what Weinstein in Chapter 2, following Klein (1986), refers to as "dynamic flexibility". Flexibility is an over-used term in discussions of the distinguishing characteristics of contemporary industrial organization, in contrast to those of previous production regimes. The fact that it crops up so often in both academic and normal discourse signifies its pervasive presence if not its academic rigour. However, Weinstein's analysis goes some way to redressing the latter weakness.

Weinstein contrasts "static flexibility", seen as the simple capacity of a firm to adjust its current product-mix and production technology to adapt to changing demand, with "dynamic flexibility", which is a far more transformative concept and practice. A dynamically flexible firm is not only capable of meeting short-term market fluctuations but is able to change its framework of operations to absorb and optimize the value embedded in innovative technologies. Organizational change and technological change are intertwined under conditions of dynamic flexibility. In particular, the existence of a high degree of responsiveness on the part of the organizational and technological departments is a defining characteristic of dynamically flexible production. In exploring this relationship in C&C industries the book, and in particular Chapter 2, assists considerably in our understanding of more macroeconomic processes of adjustment, such as those addressed in the regulation school of economic thought.

Regulatory institutions and the networking propensity

The regulation school macro-perspective is used in this book as a "critical heuristic": a model that orientates the research process (and will be tested against the empirical reality of the computing and communications industries). Theoretical attention is usefully drawn to both matters of economic organization and appropriate institutional frameworks. This is done by the conceptual devices of the "regime of accumulation" and the "mode of regulation". As may be seen in Chapter 3, there is widespread interest in the extent to which a transition is occurring in both the broad accumulation regime and regulatory mode by which the economic development process in the advanced economies is organized.

The regulationist perspective postulates that over lengthy periods of time the accumulation regime and the mode of regulation come into a reasonably synchronized relationship. The ways in which businesses are organized, their forms of work organization, the technology used, their management practices and their posture towards markets – display, despite variations at the micro-level, certain recognizable broad characteristics. Moreover, the institutional framework, which envelops and pervades key aspects of business activity, itself takes a recognizable general form across national economies. The kinds of state intervention in support of economic activity (monetary, budgetary and fiscal arrangements); the degree and kind of deployment of policy instruments designed to support businesses; the rôle of informal regulatory institutions such as business associations, industrial relations practices and local chambers of commerce; and even the expectations of workers and consumers – all come to share some common and complementary features. There will also be elements of dissonance present in these relationships; the system is by no means closed.

A key claim of the work informed by this broad perspective is that, since the 1970s, one set of relationships between the accumulation regime and the regulatory mode has been discernibly yielding ground to a new set of such relationships. It is proposed that the era from approximately the 1920s to the 1970s in the USA, and the 1940s to the 1970s in Europe, can be characterized as Fordist. That is, the predominant model – not everywhere implemented, but perceived as a highly modern, impressively productive system to be aspired to – was one that originated in the eponymous American automobile firm. Among the key, and perhaps idealized, features of that micro-regime of accumulation were: flow-line assembly, detailed division of labour through task-specification for workers, mass-production of relatively standardized products, a bureaucratic divisional management hierarchy, and a market disposition based on the idea of stand-alone competition.

Crucially, what links the micro-regime of accumulation to the micro-regulatory mode under Fordism is wage bargaining. To move towards

6

synchronization of mass production with mass consumption, a relatively large increase in wage levels is necessitated. This is ultimately secured through recognition of trade unions for bargaining purposes, but the first step is taken at the micro-level by the firm – historically the Ford Motor Company in this case – raising wages substantially and setting a target which later becomes a standard for other firms to meet. In time, this linkage between mass production and mass consumption in the economy at large becomes generalized. Keynesian demand-management regulates the amount of credit in the economy, institutions of collective bargaining become established, and the welfare state develops. Economic relationships become subject to forms of regulatory institution.

Now, taking this historically situated case of the working-out of the relationships postulated in the regulationist model, to what extent does it work for the key industries explored in this study? Clearly, the answers to that question cannot be prejudged, and will, it is to be hoped, emerge in the chapters that follow. Nevertheless, some pointers can be suggested at this introductory stage, given the published research on, particularly, the histories of the telecommunications and computer equipment industries (the C&C services industry is too young to have a history).

As Hills (1984) shows, the early history of telecommunications in many countries resulted in a small number of large companies already involved in some aspect of electrical or telegraphic production and/or experimentation coming to dominate the market reasonably early. Such was the domination of the Bell Telephone Company that its international activities (taken over by ITT) had to be divested early as a consequence of anti-trust legislation in the USA. In Britain and France a very small group of preferred suppliers came to dominate (initially semi-protected in Britain by the state through the Bulk Supply Agreement). In France, Ericsson and CGTT-ITT performed a similar rôle. In both cases an almost classic Keynesian mixed-economy relationship existed between customer and supplier. That is, the public sector owned the service and undertook R&D, then essentially gave its R&D and experimentation results to the suppliers to work up first as prototypes, then final products. The state was thus bearing the high costs of research while the private sector was reaping profits from a quasi-cartellized set of supply arrangements. This does not sound Fordist in the sense that the early development of the automobile industry in the USA does. In some senses it appears almost as a form of hyper-Fordism, where regulation penetrates and almost substitutes for (UK and France) or creates (USA) the market.

Production of telecommunications equipment includes the classically

7

Fordist assembly of telephone handsets, often carried out in branch plants in regionally assisted, former heavy-industry areas to where late-Fordist industry often migrated (see, for example, Massey & Meegan 1978). However it also includes non-Fordist production such as cabling, with the case of switchgear being a more customized craft and assembly production process. One feature common in the telecommunications equipment industry is the high level of trade union representation found there in Britain and France. So, in general outline, the telecommunications equipment industry has some unique features, notably a history of close government involvement in its organization and, as a key customer, its markets, with others that could be said to resemble aspects of Fordist production technology. However, by virtue of the technology involved, not all of the industry could be compared to, for example, the automobile industry.

Computing, perhaps, can be so compared in that much of the production of computers involves assembly of electronic components and printed circuit boards on assembly lines and, as the industry has moved more and more towards producing a mass commodity rather than a crafted, one-off item, this is becoming increasingly the case. Offshore production of components and assembly of final product is extremely common, as low-wage platforms have been sought for this labour-intensive process. Like the automobile industry, a large number of early producers gave way to a dominant market-leader (IBM in the USA, ICL in Britain, Bull eventually in France) who was pursued for a time, especially in the USA, by smaller imitators, the numbers of which have now reduced to a handful (and in Britain and France to negligible proportions).

In both Britain and France the state has at various times intervened significantly to restructure the computer industry. In France, even today the main player, Bull, is hugely subsidized by the state through its holding company, as Chapter 6 makes clear. In Britain, the history of ICL is one which is indistinguishable from state involvement, since it was the Labour government, through its Industrial Reorganization Corporation in the 1960s, that gave birth to the company, taking the computer divisions of a number of electronics companies to form ICL. Thereafter, ICL's difficulties in trying, and failing, to compete with IBM have often drawn forth special government assistance in the form of both bail-outs and government purchasing policies. Computer companies tend, however, to be non-unionized, following their US exemplars, and wage-bargaining can be undertaken on the basis of individual, let alone local, contracts.

Thus, from this brief and rapid overview some conclusions may be drawn. First, both computing and communications display characteristics to varying degrees that may be characterized as Fordist. Secondly, for reasons having to do with the strategic nature of the technology, telecommunications looks historically to have been an almost hyper-Fordist industry, given the degree of state regulation that enshrouded its production capability. Finally, though, the industries, hypothetically at least, differ in their industrial relations cultures: in telecommunications equipment, unions are more common and powerful than in the computer industry.

If the question of whether or not the two industries can be truly characterized as Fordist can be established only by detailed historical analysis, nevertheless another more germane question arises. If, as the regulationist perspective suggests, Fordism is declining to be replaced by a newer regime of accumulation and mode of regulation, what general configuration do these now display, and is there evidence that C&C industries are being transformed accordingly? In Chapters 6, 7 and 8 these issues are tackled as a major theme of the book.

Put simply, Chapter 3 suggests, along with common parlance in the appropriate literature, that we are entering a post-Fordist economic regime with an associated – broadly post-Keynesian – mode of regulation. The term post-Fordism is something of a convenience. The forms of business organization which have been argued to be post-Fordist include the following: substantial dependence upon networks of suppliers and the elaboration of supply-chains; intensive use of programmable assembly, facilitating a high degree of production flexibility; matrix-like and more decentralized, less bureaucratic, management structures; higher skill-densities in workforces; more flexible working practices than hitherto; and a heightened tendency towards collaborative R&D, technology transfer and marketing arrangements among firms.

At the heart of post-Fordism is a greater degree of *integration* inside and between firms, and in terms of functions and organization. To some extent, integration may also extend beyond the firm into the micro-regulatory sphere; that is, there may be greater interaction with the institutional context of local and regional as well as larger-scale governmental bodies, business associations, chambers of commerce, training providers, universities, and so on. The success of such practices depends crucially on the quality of *networking* undertaken by firms and their institutional support systems. Communications networks – the very technologies produced by the computing and communications industries – are of central importance to the efficient and effective functioning of business networks. Firms in these two sectors are often leading-edge users as well as suppliers of such

9

expertise, as Chapter 8 shows. The information and communications network infrastructure is what enables but also demands the flexibility supposedly characteristic of the post-Fordist firm.

Globalization and corporate strategy

The growth of the world economy is presently accompanied by the diversification of the corporate origins of the businesses responsible for global economic expansion. In the relatively recent past, the internationalization of investment, as distinct from trade, was stimulated by large corporations operating transnationally from a few national economies (the USA, UK and some other European countries). Now it is not uncommon for relatively insignificant countries (in economic terms) to be the originators of what can be called global corporations: Switzerland's pharmaceuticals and food industry giants come to mind, as do South Korean electronics and automotive firms, for example. Moreover, a proportionately greater share of such corporations originate from outside the USA than has historically been the case. In the words of the Japanese business-strategy analyst Ohmae (1985), such corporations are seeking to be as fully economically active in the centres of "Triad power" as possible, the Triad in question being the three main global markets of the USA, Europe and Japan.

These developments are the subject of Chapter 4 of this book, which examines the growth of global economic investment activity and market penetration, not only with reference to the information industries that are the focus of the book but to others, such as financial services, which are the most globalized of all. Despite the ubiquity of financial services within and beyond the "Triad" markets, stimulating the development of important financial centres in developing-country cities such as Bombay, São Paulo and Nairobi, computing and communications are certainly highly internationalized industries well on the way to becoming global. That is, no longer do American firms dominate world markets supremely, as they used to do. Rather, major challengers have arisen in Japan and, to a lesser extent Europe – particularly in the case of computing, where indigenous firms are weak. Some Japanese firms have the strategic aim of, for example, outperforming IBM in computing, as Fujitsu have gone on record as saying, and AT&T in communications, as is the case with NEC. European firms such as Siemens in computing and communications, and Alcatel in communications, have professed aims of becoming, or consolidating their

10

positions as, global corporations. This means not only selling to US and Japanese markets, but also establishing production, design and research facilities in those markets. There are, of course, considerable political difficulties and obstacles in achieving this. But there are also ways round them, and firms with global intent have been learning a considerable amount about these in the past decade or so.

One of the key reasons globalization has begun to occur is precisely because of the progress made by computing and communications technologies themselves. These enable corporations to deepen their vertical structures of production in host countries and to interchange what may be their own internal technical, production or marketing complementarities across country and continental boundaries. The case of IBM is instructive in this respect. That corporation was organized such that by the 1970s it had distributed fundamental research centres in the USA and Europe, and a multitude of development laboratories throughout the USA, Europe and in Japan. However, there was no necessary downstream complementarity in the production specialisms that might also coincidentally be located in particular countries or even continents. Intra-corporate electronic networking provided the crucial highways down which such complex functional and spatial interactions could be co-ordinated. Of course, IBM has long played the part of "corporate exemplar": other companies observe and learn from the best practice of the world's leading computer manufacturer; thus the experience has slowly become more generalized.

However, one of the open questions about globalization, which this research was committed to addressing, concerned the extent to which a producer and production-led IBM-style model of global supremacy still pertained. As will be recalled from the previous section, theory suggested that the era of Fordist producer-power might well be expected to be under challenge from new models of industrial organization post-dating Fordist production. As will be seen in Chapter 6, IBM – along with its hot pursuers DEC, Hewlett–Packard and the like in computing – had been assailed by what can only be called the decline of producer-power and the rise of user-power. The demands of the market, with heightened competition, appear to have pushed even the powerful IBM down developmental pathways which cause them to build upon and extend their networking competences in new ways. All computer and communications companies are prey to the force of the global market pulling them out of their national bases, both to learn from the innovation specialities of technologists in other countries and to meet an ever-discerning customer demand.

As well as significant changes in markets, and developments in

11

technology facilitating the management of greater global reach, two other stimuli to globalization are of interest to this research. These devolve into questions about the rôle of government regulation and of corporate collaboration. The question of industry regulation has been of greatest significance in telecommunications service delivery (outside the main ambit of this book) but directly in the equipment industry too. The "regulation school" perspective deduces a likely change in the general regulatory environment in association with the decay of a specific regulated accumulation regime such as Fordism–Keynesianism. In no industry has this shift been experienced more pointedly than in telecommunications where, to help restore the profit-making capacity of industry faced with the general crisis of the 1970s, governments in two of the most clearly weakening advanced economies, the US and UK, deregulated and privatized telecommunications respectively. Hypothetically, this opened up domestic markets, forcing national champions to seek sales elsewhere, a globalizing tendency the detailed implications of which are adumbrated in Chapter 4 and pursued in Chapter 7.

Finally, in the face of all the foregoing uncertainties, how, it logically follows, were businesses responding? As noted already, globalization might be a post-Fordist regime tendency, but there remain many corporate and state-regulatory obstacles to its achievement. In Chapters 6 and 7 particularly, it is shown that, not only have C&C firms been among the earliest to experiment with macro-networking based on inter-corporate collaboration and partnership, they have accordingly been among the first industries to learn about the pitfalls of partnership and how to deal with and even transcend them.

Spatial change at the meso- level

The information technology industries, of which computing and communications form the core, have been widely seen as having an important rôle in regional development. This is because it is believed that they have major job-generation potential, as studies of employment growth in, particularly, US high-technology complexes have shown (for example, see Scott 1988). The reality in Britain and France is more mixed and, while there have been regional shifts in both employment and occupational structure (as Ch. 5 shows in some detail), they have been massively outweighed by the continuing dominance of the metropolitan regions of each country as major centres of employment.

Looking briefly at the *national* picture for Britain and France first, the 1980s displayed more subsectoral employment decline than growth. This was particularly pronounced in Britain where, in hardware sectors, only electronic data-processing equipment, office machinery and consumer goods showed employment growth between 1981 and 1987. In France, where the subsectoral classification is more detailed, the largest employment increase was in industrial automation, with semiconductor production and electronic data-processing equipment also showing increases. In both countries, computer services grew substantially as sources of employment.

Broadly comparable *regional* employment shifts can be identified in the two countries. The most striking change during the 1980s was the sharp employment decline in the two capital cities of London and Paris, accompanied by some sectorally variable employment growth in their surrounding metropolitan regions. London experienced a catastrophic decline in jobs in telecommunications equipment, from approximately 12,000 to 2,000 between 1981 and 1987. Much of that loss occurred between 1984 and 1987. London also lost employment in electronic data-processing equipment, but to a much less significant degree. Greater Paris showed an employment decline in the more broadly defined information equipment industry from around 140,000 in 1981 to around 117,000 in 1987.

The regions showing the larger employment increases tended to be at a distance from the metropolitan centres. In Britain, Scotland, Wales and the West Midlands increased their share of electronic data-processing jobs the most, while in a universally declining telecommunications equipment industry the West Midlands, Wales and North West England increased their shares of employment. In France, the leading employment growth regions in the information equipment industry were in the south (Rhône–Alpes, Midi–Pyrénées, Provence–Côte d'Azur) and Alsace. Each country thus displays an emergent though highly skewed regional pattern of employment in these industries. This is characterized by a strong metropolitan region (with a weakening centre) and one or two growing outliers, notably Scotland as Britain's second electronic data-processing region and the West Midlands for telecommunications equipment. In France the comparable sectoral/regional split would place Rhône–Alpes as the second computing region and Brittany as the second telecommunications equipment region.

For information technology services the picture is different, both from that for hardware and for Britain as compared to France. Metropolitan centres dominate the location of these activities in both countries. However, in France there is a marked and growing concentration of this activity – on a far lesser scale – in Rhône–Alpes and Provence–Côte d'Azur. In Britain there have been quite significant changes during the 1980s. In the early part of the decade there was employment decline in cities, including London, and a spreading out of employment growth into less urbanized areas. After 1984 this trend reversed quite markedly, and growth was more concentrated in South East England, with London maintaining an unchanged employment level. In telecommunications services the early 1980s saw patchy growth and decline in employment throughout the southern half of Britain, while the late 1980s were marked by widespread employment decline except in parts of the outer South East and Scotland.

In theoretical terms what seems to have been happening is consistent with analyses that postulate a Fordist spatial division of labour in industries such as computing and communications hardware (Massey & Meegan 1978). That is, production functions leave headquarters locations, possibly to be replaced by managerial functions in city locations, especially in the metropolis. Restructuring also affects employment in traditional manufacturing regions, as firms shed labour to meet competitive threats from foreign companies. Inward investors and new domestic investments may be made in perimeter regions (due to labour availability and government investment grants) or in outer metropolitan areas where access to highly qualified professional and scientific labour is prized.

A key question in light of the regulationist analysis -- positing as it does the prospect of a transition from older Fordist structures to something else – is whether the regional shifts point to new kinds of employment? Fordist industry requires low-skill assembly labour in substantial numbers; post-Fordist industry is said to require higher skills. The evidence in Chapter 6 suggests that the decline of Fordism is indeed accompanied by the slow decline of the less-skilled occupational base and a rise in the share of higher-qualified workers. In Britain, from 1983 to 1989 the electronic data-processing occupational structure shows a noticeable gain in the share of professionals at the expense of operators in all regions except East Anglia, the East Midlands and Scotland. Scotland is the anomaly, and it seems likely that Fordist tendencies still persist in the Scottish computer transplants. The largest relative growth of professionals is found in the North West (the main domestic production base in UK computing), the South East and South West England. The position in telecommunications equipment production is similar but far more dramatic. All regions show a growing share of professionals in this industry; all regions show a decline in the operator share. Some regions such as the North West display an

14

enormous switch in these occupational ratios. Operator grades continue to predominate the further north and west one moves away from southern England.

The position in France is comparable. In Paris the share of engineers in information industries doubled between around 1979 and 1987. All French regions display, from a much smaller employment base, the same or an even more sharply increasing share. Non-qualified employees, by contrast, are practically everywhere a declining proportion of the work-force. The regions where particular occupational shifts reinforce a spatial division of labour are the southeastern (Rhône–Alpes, Provence, etc.), where there are higher proportions of engineers and technicians, and the western (for example, Brittany) where the opposite had been the case but where a switching almost comparable to that of North West England occurred in the 1980s.

The dynamics of the C&C industries

The computing and communications equipment and services industries are highly dynamic in terms of the development of markets, technologies and corporate organization. Even though some of the world's largest firms such as IBM and AT&T are prominent in these sectors, they have by no means been of a sufficient scale to enjoy immunity from the forces responsible for change throughout the industries. Indeed, such has been the volatility that precisely these largest of corporate actors have been placed under challenge in fields over which they hitherto held sway. Mistakes of a major and expensive kind have been made as they sought to develop cross-sectoral competences or failed to develop organizationally in the face of the new networking dynamic. Other corporate giants, notably Philips, have been placed in serious jeopardy by their failure to come to terms with the new competitive realities.

Perhaps the key organizational and technological force placing the established "M-form" corporations (Chandler 1977) in difficulty – with their hierarchical, bureaucratic structures of administration, management of innovation and marketing – has been their profound lack of flexibility. Slow in developing new products, not least because of their leviathan structures, they have been to some extent outwitted by new, more nimble market entrants challenging not with newer or better versions of their competitors' products but with new concepts of how the product should be designed and marketed.

The competitive struggle in computing and communications has been similar, as Chapters 6 and 7 show. It is between hierarchical and decentralized firms. The traditional market leaders in C&C produced large-scale "installed base" hardware at high cost for large customers: government departments, large corporations, universities, etc. In computing this was the era of the large mainframe product, the market in which IBM came to dominate on a world scale. In communications the public switch, the centre of a large-scale network, built for large public or private customers, played an equivalent role. Barriers to entry into these markets were astronomically high, so producers of these products enjoyed effective monopoly.

However, new and perhaps more democratic ideas of how to make C&C power available to a larger market, hitherto untouched by the hierarchical producers, led to the miniaturization of both computing and, for special uses, communications hardware. A new market was created with the advent of mini-computers capable of being installed at departmental rather than corporate headquarters level. The demand for more localized communications capacity, and indeed alternative transmission systems to those controlled by the big public switch, led to a similar decentralization trend in telecommunications equipment. New market entrants, and some older companies that could no longer compete with the giants, came to dominate these niches. Interestingly, the niches grew during the 1980s to challenge the dominance of the hierarchical products, thus transforming the market. Most computing sales are now at the smaller, personal-computer end of the market, though established technologies and firms still dominate the telecommunications market.

These changes have created opportunities and enormous market potential for the provision of services to C&C users. The computer services industry is now the most dynamic growth area in the C&C field. This is so much the case that a very large proportion of the revenue of the large hardware producers is now derived from the process of "systems integration", the installation and adaptation of hardware configurations, the writing of appropriate software, and the customer aftercare and debugging associated with the spreading of C&C competence into businesses that, in the past, would not have been able to afford such facilities. However, the hardware producers are by no means alone in recognizing this market. Software and systems-engineering companies, specialist consultancies and the global accounting and management consultancies are all active in supplying systems know-how to user-firms.

There are interesting contrasts between countries in the nature of this provision, as Chapter 8 makes clear. In France, as with many other continental European countries, the large Anglo-American management consultancies have a major share of the market. In Britain, which has for some time had a good reputation for innovative software and systems engineering, firms in those sectors have been major providers of computing and communications services consultancy. However, that situation is itself changing as relatively undercapitalized and labour-intensive software specialists recognize the need for investment capital from larger firms in order to expand their scale economies and, to assist this, invest in computer-assisted systems for software production. Hence in the late 1980s and early 1990s there has been a spate of acquisitions, mergers and alliances as concentration of the industry proceeds apace. French software and systems houses concentrated earlier than those in Britain, and there has been acquisition activity from that direction as well as from, for example, AT&T, who acquired the UK firm Istel.

Thus, in the three main spheres of computer equipment, telecommunications equipment and C&C services, there has been a considerable transformation of the products and services provided. Markets have expanded and deepened as costs have come down, technologies have become more user-friendly and flexible, and more and more businesses have become aware of the imperative of being networked into their markets. Firms have become more reliant upon chosen others in order to maximize competitive advantage. Competition has been heightened by the rise of new competitor nations (notably Japan), technological innovation, and market expansion. Groups of companies have engaged in a range of collaborative activities in order to reap advantage from competitors at least cost. The particular spheres in which such competition by collaboration or acquisition have been most pronounced have been in R&D, technology acquisition and to gain access rapidly to new markets (or seek to protect old ones).

In some of the leading cases, government activity has been important in setting the agenda for competition. In Britain, the privatization of British Telecom opened up the equipment supply market, weakening hitherto protected domestic suppliers and forcing them to pursue more internationally minded strategies. It also allowed overseas firms to become active in the domestic market. In France, government policy towards both computing and communications has been crucial. Without government aid, Bull would not be the global player it currently professes to be. In telecommunications, France Telecom remains publicly owned but there has been some, albeit French-dominated, liberalization of the equipment supply industry. Clearly, the relative importance of state regulatory positions must not be down-played in considering the reasons for the contrasting performance of these industries as between Britain and France. However, it would be equally misguided not to recognize that beyond the boundaries of these particular nation-states, global forces of technological change, competition and market expansion are structuring the space that national governments have within which to manoeuvre. This proposition suggests that, presently at least, sectoral dynamics are of primary importance in determining the trajectories of these industries, and that for the moment national governments and supra-national organizations such as the European Community must grapple with these as best they can.

CHAPTER TWO
High technology and flexibility

Olivier Weinstein

This chapter raises some questions on the nature of contemporary transformations of technologies and productive structures and, at a higher level of analysis, on the emergence of a new mode of industrial organization or a new "productive order". These questions follow from a confrontation of two lines of reflection.

The first one starts from an analysis of the crisis and the transformations of mass production on the Fordist model. This model is said to be the basis of the long expansion stages of capitalist economies. This approach stresses the shift to flexible technologies and production systems, which, according to different views, will lead to a substitute for or a prolongation of the mass-production model. The second line of reflection is rooted in a view of the development of high-technology industries, which one can also qualify as "science-based" industries; in particular, C&C industries. Reference to these industries can be justified by the special place they occupy in the reconstitution of productive systems, but this justification bears on the idea that they are carriers of the most fundamental transformations of the productive structure, both as locus of emerging new forms of industrial organization and by their impact on the entire productive system. This impact is produced through the diffusion of new production trends, new means of consumption of new technologies and knowledge and, more broadly speaking, new perspectives.

In this way, one is confronted with two views of technological change which can develop separately and offer significantly different lines of analysis. This dichotomy becomes clear when referring to the modes of development of different industries since the end of the Second World War. But it can also contain, in a more or less explicit form, different theorizing on technological dynamics and on the evolutionary and transformative *forms* of production processes and industrial structures.

In this chapter, we want to examine the extent to which it is possible to make a synthesis of both approaches. We seek to do so by clarifying different notions of flexibility, and by seeking to account for what could be designated a new "productive paradigm".

Revisiting post-war economic growth

For a better understanding of contemporary transformations in productive structures, it is useful to revisit certain aspects of post-war growth and crisis, notably from the point of view of the evolution of productive systems. Without being able to examine this in detail, we propose a number of hypotheses regarding the evolution of two trajectories that were originally relatively autonomous, but which presently display progressive convergence.

Post-war growth: two parallel movements

In a simplified way we can say that the long post-war phase of accumulation of capital rested on two axes of industrial development, connected to two "productive forms" (Weinstein 1983). First is a movement of intensification and generalization of principles of mechanization, according to the "Fordist model". This central dimension of the development of capitalism since the beginning of the century, which found its origin in the "American manufacturing system", has been the object of many analyses (Rosenberg 1976). In essence, one can provisionally describe this as a *mass-production system*. But, in parallel, industrial evolution has been characterized by new activities which, originally, developed independently of the Fordist system. Key features of these activities are their strong links with science and technology, and the importance of research and development institutions in the development process. The activities can be considered as the key locus of emerging and actual new industrial dynamics and technological trajectories.

Although both productive forms can occur in the same sector, or in the

20

same production process, they vary according to the dynamics of different industries. In this way, post-war growth is led by two types of industries (Freeman et al. 1982):

(a)   mass consumption industries: automobile and other consumer durables typical of the Fordist model;
(b)   high-technology intensive industries: chemicals, electrical and electronic equipment, telecommunications and instrumentation.

The elements of a new industrial model have been fostered especially in the second group, to which one should add aerospace as well as the entire military equipment sector. The C&C industries have a privileged position and they are the main generators of the factors underlying a global transformation of the technical system.

The development of these new activities initially occurred autonomously, relatively independent of the logic of the Fordist system – the articulation between mass consumption and mass production – according to a specific process with three dimensions:

(a)   a technological trajectory characterized by continued bursts of combined product and process innovation;
(b)   new forms of industrial organization within and between firms, such as the development of flexible production complexes of the kind found in Silicon Valley;
(c)   innovative forms of demand often linked to the supply of innovative products stimulated by military and civilian public-sector procurement; in the latter case, aerospace and telecommunications technologies are key examples.

The process of transformation of these new industries, as well as their growing importance in the productive system, play a major part in restructuring movements of productive structures.

Which convergence in time of crisis?
Following this line of reasoning, we formulate the hypothesis that crisis and contemporary transformation of the productive system have a double dimension. On one side, a crisis of the mass-production system, i.e. a crisis of the Fordist production process. This crisis followed from the exhaustion of its efficacy, particularly regarding the twin pressures of technological and labour-process change. Yet it is fed by the evolution of

21

the nature of demand and the conditions of competition. From this point of view, the present transformations are characterized as a shift from the Fordist system to a new industrial model of "flexibility". On the other side, the rise of new, high-technology industries as carriers of new industrial organizational forms. This change disintegrates and transforms the whole of the productive system, and focuses the analysis of restructuring on the links between research and production, the new dynamics of technological change and the emergence of a so-called "knowledge-intensive" economy.

Major theoretical and empirical questions regarding contemporary transformations of production stem from these two dimensions. However, there is some convergence between them caused by the diffusion of new technologies, especially C&C technologies, and more fundamentally by a transformation of industrial logic in the sense of a generalization of the "high-technology" model. This generalization also encompasses a "de-maturing process" of traditional industries (Abernathy et al. 1983).

The economy of flexible systems: some questions

Flexibility as a new industrial model

Identifying the main characteristics of the transformation of productive structures is difficult. On the one hand a crisis and restructuring phase is a moment of social experimentation; thus, multiple, often ephemeral, institutional forms appear. On the other hand, because of the constraints inherent in a long period of relative stagnation, which raises specific forms of adaptation, behaviours and modes of organization, it is difficult to distinguish between transitory and durable responses. Nevertheless, it remains possible and necessary to hypothesize about the emergence of a new model of industrial development.

A first thesis argues that the Fordist model gives way to a flexible industrial model (cf. Piore & Sabel 1984, Boyer & Coriat 1986, Kristensen 1986, Cooke 1988). In a simplified way, one can confront both models in the following way. The Fordist model, the basis of post-war growth, would rest on:

(a)   as production management principles, the maximization of quantities and the minimization of costs by productivity increases and intensification of blue-collar work; and

(b)   an industrial efficacy based on a Taylorist organization of work, the

22

development of rigid mechanization using specialized machines, the effect of real economies and the effects of learning processes through the mass production of standardized goods.

The flexible model emerges in response to the blocks resulting from the increasing technical and social rigidity of the mass-production system and, above all, the evolution of demand towards a growing instability, diversification and sophistication within the context of new competitive conditions. Technological flexibility or, to use an even more relevant term, productive flexibility, can in very general terms be defined as the capacity of a productive system to adapt itself at the lowest cost to different contingencies, in particular to both quantitative and qualitative variations in demand. Flexible production systems would have:

(a)   as a final objective: improved product quality and diversity, batch production, the development of supply-chains for goods and services, adaptation of products to the specific needs of users, with a particular dominance of small- and medium-sized batch production; and

(b)   as a means of implementation: the establishment of new forms of production and management based on new capital equipment, computerized information systems, communication systems, programmable automation, and new forms of organization which can be implemented at several levels (production lines, production units, units of firms or groups of firms).

The purpose of this argument is not to analyse this flexible model and its variations. Still, we want to focus on some questions concerning its content and consequences. Today, the search for flexibility presents itself as a major preoccupation at several levels and modes of organization of production. It includes organizational forms that move beyond multidivisional or static matrix structures, the push towards externalization, as well as network relationships between enterprises, the functioning of the labour market, and the transformation of regulatory structures. This widespread use of the concept of flexibility stresses the centrality of the transformation of productive structures. But, at the same time, it gives a multidimensional content to this notion. As such, it makes precarious the analysis of its content and consequences. From the point of view of the analysis of technological change and the transformation of production structures, two types of questions are important: those concerning transformation of demand and the diverse nature of the product, and those concerning links between mass and flexible production.

Evolution of demand and the nature of the product
The transformation of the nature of demand and the forms of competition are in general perceived as the main determinant of the need for flexibility. Although agreement on this general statement is easily established, it is nevertheless important to examine the precise nature of these transformations, their most fundamental dimensions, their origin and their links with global shifts in production conditions.

In a general way, we can say that flexibility relates to the links between the production process and demanded products. Several typologies of forms of flexibility have been proposed. Most of them were developed from a strongly technological point of view (see, for example, Adler 1985 or Gerwin & Leung 1980 in Coriat 1988). From the standpoint of economic analysis of the production system, some conceptual distinctions are useful.

First, there is the distinction between flexibility *vis-à-vis* production process and *vis-à-vis* the products. *Process flexibility* refers to the capacity of a production system to incorporate transformations in methods of production. This includes both a given output structure or a changing product array. This notion is essential for the study of *dynamic flexibility*, to which we will return later. *Product flexibility* refers to the capacity of a production process to adjust itself at a limited cost to the variation of the product. Both notions are closely related to the extent that, as is shown by empirical studies on production management, the characteristics of the product and the process are hard to separate. Among both notions, the analysis of *product flexibility* receives most attention when one seeks to define a flexible model in relation to the mass-production model. Still, determining its content remains rather complex, because the variability of the product can offer different dimensions. As a first approach, we distinguish three axes of output variability: (a) fluctuations in quantity produced for a given product, resulting from fluctuations in demand; (b) diversification of products and variation in the composition of output; (c) accelerated renewal of products, which lead to three forms of flexibility.

Flexibility can first of all be defined as the capacity of a production apparatus to respond to a variation in the produced quantity at a low cost. This is the conception that one finds in economic analysis, such as that of Stigler (1951). The degree of flexibility is, in this case, measured starting from the average cost curve of the firm: the flatter the curve, the less cost varies with the produced quantity, the more the technology is flexible. In this case, flexibility is especially a question of the weight of fixed overhead costs. *Ceteris paribus*, an increase in overhead costs, more specifically due to the use of particular assets, induces a rigidification of production. The search for flexibility should thus be accomplished by a search for reduced overhead costs or, more precisely, sunk costs. Note that this can be realized at the firm level by technological means, but also by organizational options such as externalization.

A second type of flexibility can, in general terms, be defined as the capacity of a production apparatus to produce different goods and to accommodate, against least cost, a variation in the output structure. These two features are not strictly equivalent; multi-product systems can be rigid (the output structure is hard to change) or flexible. Therefore, the definition of this type of flexibility combines both features. It is possible to specify the analysis by decomposing this case according to the degree of differentiation of products (e.g. by distinguishing flexibility of series, when talking of varieties of the same base product, and product flexibility in the real sense of the term; cf. Coriat 1988).

This notion of flexibility is particularly significant, in part because of the importance of diversification and multi-product output analysis in itself, but also because of the close links between flexibility problems and the development of production diversification. The need for flexibility stems in the first instance from qualitative changes in *demand* and firm *supply*. The links between flexibility and multi-production or, in a wider sense, between flexibilities and product forms, require examination. The diversification of production is not a new phenomenon, as becomes clear from Chandler's (1966) work, for example. Still, the analysis of the multi-product industry and firm has only recently been developed. It can be conducted following approaches which reflect significantly different conceptions with respect to the content and *raison d'être* of diversification and multiplication of products (see Teece 1986). Product multiplication can in fact be realized along three different trajectories, possibly corresponding to different consequences for productive structures.

The first trajectory involves the differentiation of a same-type product by multiplying models and varieties for each model. This tendency is not novel: in the automobile industry it was initiated by General Motors from the end of the 1920s. One can consider this tendency as a central dimension in the evolution of the mass production system (modified Sloan–Fordism or Sloanism). True, in Europe it developed only later and, because of the evolution of markets (in particular their internationalization) it developed to an unprecedented degree. In the automobile industry, the total number of varieties has grown from some tens at the beginning of the 1960s, to tens of thousands in the middle of the 1970s.

According to Coriat (1988) and Teece (1986), this is one of the causes of the crisis in the mass-production system which has led to the emergence of flexible production systems. It is only one aspect of the evolution in the nature of products, and not necessarily the most important.

The second trajectory requires diversification in the real sense, i.e. the firm produces and supplies a set of goods that are different in terms of their use-value and their production conditions. The borderline with the previous case is hard to determine, but for our purpose we postulate that it exists. Leaving aside the conglomerate form, diversification can follow a supply or demand logic, or both. In the first case, diversification aims at a better usage of certain assets and, in this way, seeks to benefit from *economies of scope*. It is important here to identify the different types of assets at the origin of economies of scope and, especially to identify immaterial assets – knowledge in particular. It is on this aspect, the advantages of diversification cost, that the analysis of the multi-product firm focuses.

But the diversification of the products which the firm begins to offer can follow a third trajectory, i.e. to meet the nature of certain demands. That happens according to two, closely related models: (a) the supply of sets of complementary products (goods or services), or of systems consisting of different components which strongly differ in their productivity conditions and the assets they mobilize (type of knowledge); and (b) the customization of products and systems, corresponding to the specificities of different demands, resulting in small-series or piecewise production. This is without doubt one of the strongest departures from the mass-production model.

These different diversification modalities must lead to diverse configurations of productive structures. The types of flexible technologies and organization models should be studied in this context. In this chapter, we confine ourselves to pointing out the ambivalence of the relations between flexibility and multi-production.

On the one hand, diversification can be a means to respond to the problems posed by demand fluctuations and rising overhead costs. Since the appropriate use of technologies allows variations in output structure, adequate management of the product portfolio facilitates the growth and stabilization of the capacity utilization rate, and the possibility of benefiting from scale economies. From this point of view, flexibility is approached as a mode of simultaneous application of scale and scope economies. On the other hand, flexibility necessitates facing the new demand forms mentioned earlier. This aspect is essential today, especially in high-technology industries. It can lead to specific productive forms related to the preceding case, to the extent that, for example, the search for lower costs is neither the only nor the most important preoccupation. For the firm, the problem is not just to know how to make a product or a set of products at the least cost, but also to determine which products should be made and to what specifications. Further, the problem is often not to utilize existing assets better, but to manage larger and more diverse sets of assets, to meet the complexity of demand. However, these questions can be fully understood only by deepening the analysis of forms of flexibility.

The third dimension of product flexibility concerns the capacity of a production system to assure renewal and rapid modification of products. This aspect is, in our opinion, the most important one in the analysis of contemporary transformations of product structures. The accelerated renewal of products and production conditions is one of the major features of high-technology industries, and is about to affect the whole of manufacturing, and even parts of service industry. This renewal poses specific problems for the global organization of production. This should lead to a novel approach, which stresses *dynamic* flexibility. We will develop this notion later in this chapter. For the moment, we adopt a very general definition: dynamic flexibility is the capacity of a system to transform itself by producing and assimilating innovations of different types, and to adapt rapidly to a changing environment.

Mass production and flexible production

The multi-faceted content of the notion of flexibility leads to the following question of the relationship between mass- and flexible production. Are they two radically opposed generic forms of industrial organization, with flexible production being a new industrial model, a new techno-economic paradigm, or a new regime of accumulation? In other words, is the shift from mass to flexible production at the centre of present contemporary changes in productive organization?

It is clear that the answer to such a question depends on what precisely is meant by the notion of flexibility. In fact, if one sticks for the moment to a common notion of flexible production – i.e. an organization aiming at the production in small or medium-sized series of a collection of diversified products, the content of which can vary rapidly – then this notion can be considered neither new nor opposed to the system of mass production. This becomes clear when considering some essential features of technological and industrial evolution of the Fordist stage, as well as of the development

27

of certain high-technology industries which are at the centre of contemporary transformations.

Historical analyses of growth since the beginning of this century show clearly that this growth cannot be reduced to the development of mass production of standardized goods, as could be suggested by some conceptions of Fordism. Two essential features should be retained.

First, product diversification is not new. It became important in the 1920s, taking the form of a *policy of series*. The evolution of General Motors is characteristic of this tendency, which is accompanied by a new form of organization of the enterprise: the multi-divisional form or the M form. Since Chandler (1966, 1977), the historical importance of this organizational innovation and its links with the evolution of markets and technologies are well known. Under these conditions, it is preferable to characterize the industrial system dominant until the crisis of the 1970s as a mass-production system of diversified character, a system which simultaneously applies scale and scope economies. Scope economies are sometimes presented as being specific to a flexible production system (one could call it "Sloanism", as an extension of Fordism). Contemporary transformations in mass-production sectors such as the automobile industry should be situated in this general framework, at least as far as the transformation of the manufacturing process by the use of robot technology and flexible automation are concerned. As such, these transformations push mass production even further by developing what one can call *flexible* mass-production systems.

Secondly, the development of mass production based on the diffusion of "machinism" went together with the development of specialized equipment goods. The growth of the equipment-goods sector has shown features close to those traditionally attached to the new flexible structures: production in small series of specific goods adapted to different users, predominance of small and medium-size specialized firms, privileged rôle of skilled labour. Rosenberg (1976) has shown the essential part played by the equipment-goods industries and by the close interdependencies between these industries and the rest of the productive system, especially large industry, in the constitution of an "American manufacturing system". The development of the mass-production system is, as such, inseparable from a sector with strongly different structures and technical forms. In this sense, mass production and flexible production could be interpreted as two faces of the same industrial dynamics.

The development of high-technology industries reveals a relative rather than absolute opposition between mass and flexible production. Moreover, scale economies and, even more so, learning effects remain very important (Boyer & Coriat 1986). In the computer industry new activities such as the production of standard integrated circuits are at the centre of the newly developing technical system. This process arises from mass production in the strictest sense of term. However, at the same time highly original technological dynamics are present. These are more innovative than those observed in the Fordist industries of the preceding decades. Therefore, there is coexistence and complementarity in high-technology industries between mass- and small-series production, as well as between large industrial organizations (as they already existed before contemporary restructuring) and new structures such as those in Silicon Valley. This type possibly shows similarities with that observed by Rosenberg in his analysis of the rise of machinism and the place of the machine-led industry.

On this point, it is also interesting to take up the analyses by Rosenberg (1976) of "technological convergence", i.e. the process of diffusion of few technologies and basic production processes to many industries. These analyses allow us to understand certain general evolutionary forms of industrial structures as we find them today (Pavitt 1986). One of these concerns the relationships between "capital-intensive" large firms, small specialized producers and a process of "vertical disintegration", according to a logic described by Stigler (1951) which, still following Rosenberg, tends to accompany technological convergence. This facilitates understanding of contemporary evolutionary trends and the precise nature of new flexible structures.

These reflections do not mean that there can be nothing new under the Sun and that the scene today is nothing more than a replay of what capitalism has generated in earlier periods of significant economic turbulence. They suggest that contemporary transformations cannot be reduced to an opposition between mass production and flexibility, as is, for example, expressed in Piore & Sabel's theses (1984). In order to comprehend what is really new in the observed transformations, it is useful to analyse them from another point of view, by starting off with the features specific to high-technology industries. We can return to the notion of flexibility later.

The high-technology model: knowledge-intensive production

The flexibility theme springs in the first place from the analysis of the transformation of Fordist and, more precisely, mass-consumption

industries. Observing the rise of new industries leads, at least at a first
level of analysis, to significantly different issues: the development of links
between science and production, and the growing importance of research
& development.

General features

The rise of industries with high technological intensity, referred to as
"science-based" industries (Noble 1979), is accompanied by a process
transforming the socio-economic conditions underlying the production and
circulation of technologies and, more generally, scientific and industrial
knowledge. This process manifests itself by the development of an
increasingly complex scientific and technological system, as well as by a
narrowing of the relationships between science and production. The process
began at the end of the 19th century and can be analysed as a long
movement of submission of science to capital. Within this evolution, the
Second World War meant a decisive breaking point which can be
characterized by:

(a)   a shift towards the institutionalization of "big science", including
        aspects of industrialization and exposure of scientific labour to
        market exchange;
(b)   the building-up of a scientific and technical system organized
        around industrial applications;
(c)   a significant increase in state intervention, especially in the military
        sector;
(d)   a strong and practically continuous growth of research &
        development expenditures.

As Noble shows, these changes are closely linked to the overall
transformation of industrial structures and the continuing rise of oligopoly.
This process was pointed out by Schumpeter (1942) in his analysis of the
conditions of innovation in "trustified" capital. But in the beginning, this
transformation affected the industrial apparatus in a very uneven way, and
it focused only on a limited number of industries. In these industries, one
witnesses the emergence of a new technological and industrial paradigm.
A prime feature of such a paradigm is the existence of a continuous
movement generating new products and production procedures. Productive
dynamics rest on accelerated renewal of technologies more than on the
management and improvement of direct production conditions. This
renewal entails a transformation of the structure of the global production

process and a displacement of its centre of gravity. Conceptual work, in the large sense of the term, becomes crucial, as does the organization of links between conception, fabrication and marketing (conditions of usage) of goods. Marketing determines the capacity of the productive system to ensure a continuous renewal of products and techniques.

New production conditions are expressed as changes in the structure of the work-force, characterized by a rise of highly skilled labour (engineers, technicians) and by the importance of the organization of the work process; and as a transformation of the content of capital accumulation in the form of increasing "immaterial" investments: in particular, research & development and training.

This industrial model has roots in the chemical and electrical industries. During the past thirty years, its development has centred in the electronics industry and it can be characterized by:

(a)   a technological break induced by innovations in micro-electronics (the transistor, followed by the integrated circuit) and leading towards the constitution of a new technical system;

(b)   an intensification of the technological innovation process, typical of the nature of the new technological trajectory and the evolution of competition: technological change becomes continuous, cumulative and more profound; the cycle of research, development and production is accelerated;

(c)   a widening of the process by a significant development of the application of new technologies and the diversification of their usage. As such, one witnesses a unification and an integration of technological change in very diverse domains.

All these developments in combination add to the complexity of techniques and products, leading to the creation of a dense system of technological interdependencies, strictly conditioning the evolution of corporate structures and strategies.

Knowledge and industrial structures
The transformations described above lead to a reconsideration of the mode of analysis of production processes and, in turn, of the logic at work in industrial structuration. The growing importance of knowledge – its production, circulation and accumulation – is notable in this respect. This evolution is visible at several levels (Pelata & Veltz 1985):

(a)   the growing importance of conception relative to production;
(b)   changes in production less centred on the means of production and the organization of the production process; technology less incorporated in equipment and more directly in the product;
(c)   strong interaction between production and usage; whereas in Taylorist–Fordist industry, production and usage are relatively separate, as are production and market, because of the importance to different applications of the technological adaptation process. This goes together with a growing rôle of the user in the process of learning and of the development of technologies. These aspects have been stressed by Rosenberg (1982) in his analysis of "learning by using".

An understanding of this type of productive structure encompasses the analysis of links between science and production and of the questions posed by the economics of knowledge and of research & development (Weinstein 1988). Knowledge can be treated as specific assets which become central in the definition of strategies. In this context, Teece (1986) suggests an analytical scheme which is particuarly apt, studying industrial structures and strategies on the basis of three fundamental elements. First, the conditions of exploitation of new knowledge (innovations), which depend on the nature of the technologies, the forms adopted by knowledge (level of codification) and the efficacy of the protection systems (secret and/or patent). Secondly, the conditions of combining a body of new knowledge on which the innovation of an enterprise rests, and the complementary assets which the enterprise will have to mobilize in order to exploit this innovation: complementary technologies, skills with respect to production and marketing, and material assets (such as specialized equipment). Strategies and forms of organization, and in particular the choices to be made between internalization and externalization, and the definition of alliances, will be strongly conditioned by the identification of strategic critical assets, and the means of controlling them. The third element is the process of technological evolution of an industry characterized (following Abernathy & Utterback 1978 and Dosi 1982) by two large phases: the "pre-paradigmatic" phase in which the dominant features of the technology and the product (the "dominant design") are not yet fixed; and the "paradigmatic" phase, where the dominant design is constituted, after a process of selection of norms and standards. The shift to the paradigmatic stage profoundly modifies the conditions of evolution of production and competitive forms, especially in terms of the increasing rôle of economies of scale and learning. But the accelerated renewal of products and techniques in high-technology industries in the present stage can challenge the classical representations of the life cycle of an industry, as well as the consideration of flexibility constraints to which we now return.

A provisional synthesis: the economics of permanent innovation and dynamic flexibility

The two readings of industrial transformation can, as we have seen, refer first of all to two parallel realities: restructuring of Fordist industries on the one hand, and development of high-technology industries on the other. But nowadays it is more pertinent to consider them as two complementary dimensions of a global movement. The two dimensions are unified by the tendency towards dynamic flexibility in production processes and permanent innovation in product development.

The principle of dynamic flexibility

The first dimension of the present dynamic concerns the nature of the innovative process: the establishment of an organized innovative production system which transforms the general conditions for industrial development. The dynamics and competitiveness of enterprises and productive systems rest on their capacity to assure a continuous renewal of products and procedures. We enter a system of permanent innovation.

In this context, reference to the notion of flexibility can be relevant, on condition that a strict distinction between static and dynamic flexibility be made (Klein 1984, Cohen & Zysman 1987, Coriat 1988, Cohendet & Llerena 1989) and that dynamic flexibility be retained as the central feature of new productive organizations.

Whereas static (or short-term) flexibility refers to the capacity of a productive system to adapt to changing demand conditions (level and composition), dynamic (or long-term) flexibility refers to the conditions of transformation of productive structures over time. Dynamic flexibility can, in general, be defined as the capacity of a production system to develop and/or assimilate new technologies; as such, this notion evokes the rates of change and corresponding domains of a productive structure.

To talk of dynamic flexibility implies an important change in the analysis of production and the conditions of productive efficiency. It is no longer a question of the properties of a technology or a problem of technical choice but the conditions of evolution in production within a process of continuous and endogenous technological and organizational change. This has at least two important implications: (a) production must be conceived in terms of an integrated process incorporating product innovation (i.e. research & development) and also innovation in the development of production processes; and (b) the need to adopt an explicit dynamic framework of analysis, which implies going beyond the standard methods of microeconomics, mainly in order to account for the time dimension of production processes. Several conceptions of dynamic flexibility are possible. For the purpose of this chapter, we deal with two that are complementary.

A first definition given by Klein (1986), taken over and developed by Cohen & Zysman (1987) and Coriat (1988), makes dynamic flexibility a principle that aims primarily at increasing the rhythm of technological change:

"Dynamic flexibility . . . is concerned with designing production lines in a way that they can quickly evolve in response to changes in either the product or production technology. . . . The main purpose of dynamic flexibility . . . is to make rapid changes in production technology for the purpose of lowering costs and thereby improving productivity."

In Klein's work this concept originates from a comparison of the behaviour and structure of American and Japanese car firms. The merit of the concept is that it discerns different possible forms of transformation of productive structures in more detail, and it offers an essential insight into the relations between flexibility and functional organization of production. Still, this notion of dynamic flexibility remains limited in relation to the character of contemporary transformations: on the one hand it considers transformations at the level of production lines only, and on the other it limits the search for cost reductions to production costs.

Another approach to dynamic flexibility is given by Cohendet & Llerena (1989). Its context is very different, i.e. that of decision theory under uncertainty. Its formulation originates from the work of Hart (1945), who defines flexibility on the basis of the relationships between the chronology of decisions and sequences of information: a "programme of decisions" will be the more flexible to the extent that it can take into account information which is sufficiently encompassing at each stage of the decision-making process; a decision-making programme will be the more flexible, the more the information considered at the moment of decision-making is important (Hart 1945).

For our purposes, we shall retain only one definition of dynamic flexibility following this approach, starting from the possibly variable state of one production system and its associated decisions at time t, $t+1$, $t+2$, and so on, as a function of the state of the system and the decisions made in the preceding periods. The larger the domain, the more important will be dynamic flexibility. In this respect, the dynamic flexibility of a system contrasts to its irreversibility. In more general terms: dynamic flexibility should express itself in a reduction of costs and of delays of transformation and adaptation of the productive structure to the evolution of products and processes.

This concept takes the contemporary conditions of productive strategies well into account. It is not only a matter of finding the appropriate technical and organizational forms assuring a strong technological dynamism, but equally well these forms should at each stage offer diversified development potentials allowing adaptation to evolving demand, markets and technologies. This analysis then reveals its full value when one considers the different levels at which flexibility can be understood.

From the management of production lines to the management of technologies and knowledge

Dynamic flexibility as conceived by Klein (1986) corresponds to the capacity of production lines to transform themselves rapidly, so that they adapt to a continuous flow of improvement of techniques and products. This aspect is clarified by considering the factors underlying the efficacy of Japanese car firms. According to Klein, a comparison with American firms shows the rôles played by: (a) the functional organization of production lines, characterized by a strict and continuous co-ordination between departments: where design of the product, management and perfecting of equipment ("tooling" department) and production management are in constant interaction; and (b) the mode of organization of work, marked by the rotation of work posts, a strong internal mobility, and formalized relations between production workers and managers.

This would be the basis of a capacity for continued improvement of production methods and for the rapid transfer of new products and procedures, going from conception to production without high reorganization costs. In this way, dynamic flexibility assures an interaction between the evolution of products and processes based on a strong functional integration. The organization of linkages through subcontracting goes in the same direction.

Yet this Japanese example, however important it may be in analysing the transformation of mass-production industries, can account only partially for the content and consequences of dynamic flexibility, as they appear in the new technologically intense industries.

The transformation of productive structures effectively transcends a reorganization of production lines relative to a type of product, and beyond a simple change in the immediate production conditions (the "manufacturing"). Rather, such transformation concerns in a fundamental way the process of conception and the links between conception and production, and the management of technological knowledge and the relation between technologies, products and markets. To understand these transformations, it is useful to distinguish two levels which, in a first approximation, correspond to two types of industries: production and scale-intensive, and high-technology.

In the case of "production and scale-intensive" industries (Pavitt 1984), referring to essentially classic mass-production industries such as cars, dynamic flexibility remains concentrated on mastering production lines defined on the basis of well specified product types. In this sector, dynamic flexibility aims to improve the rhythm of optimizing production methods and renewal of products. It is marked by:

(a)   a growing importance of conception, and of the links between conception and production, expressed in the evolution of the cost structure;

(b)   a greater complexity of technologies and products, in the case of cars, for example, in terms of materials and components (increasing utilization of electronics);

(c)   links with a development of competition through the quality and the features of the products.

High-technology industries ("science based", according to Pavitt's terminology) witness more radical breaks with the classic forms of industrial organization. The process of technological change and transformation of production technologies is no longer centred around the management of products (or series of products) and determined production lines, but on *the global management of a coherent ensemble of technologies, competences and knowledges* (the "professions" of the enterprise), the development of that ensemble, and its exploitation on the basis of products and market spaces which are in constant evolution.

Within this context, the economy of the firm is orientated towards supply-connected projects, defined on the one hand by a permanent tension between the search for sustained accumulation of competences, modes of management of technologies and combination of technologies, and on the other by short-run constraints of feasibility and exploitability (cf. Kristensen 1986).

The capacity for permanent innovation and dynamic flexibility will rest on the general conditions of organization of technologies, products and markets at three levels. First, the capacity for mastering technology in the proper sense: the capacity to develop a technological base; mastering basic technologies, relatively simplified technologies, partially acquired outside the firm and the sector (purchases of components and equipment goods, transfer of technologies) and especially mastery of specific technologies on which the firm will base its proper domain of competence and its competitive position. This is as much a matter of producing one's own new technologies as of following and assimilating certain scientific and technological developments, as well as different types of techniques. Secondly, the capacity to utilize and combine these techniques and other assets, to conceive and produce different types of commodities (components, subsystems, final goods, services) within short time periods. This capacity for fast development and industrialization of products becomes decisive. And, finally, the capacity to react to different types of demand by offering combinations of products, i.e. material goods and often joint services as well (e.g. hardware, software and training), "complexes of commodities" or "systems merchandise" which are customized yet able to combine standard elements and elements conceived on the demand side. Borrowing an expression from industrial circles: the issue is no longer to offer products but "solutions".

The productive structures tend to reconstitute themselves around these constraints, imposed by the combination of permanent innovation and dynamic flexibility. The production of systematic and in-house knowledge of complex technologies in relation to market-demand has the following two organizational effects:

(a)   Transformation of the vertical functional structure of the production process, and more in particular of the research—development—production—usage processes, with respect to the importance of components and their modes of relationship. Two aspects are essential here: the significant rise of the rôle of research & development noted earlier, and the necessity of narrow and continuous links between conception, production and usage of the new goods. The transformation of the vertical structure also concerns the mode of functioning of the production processes: the technological complexity of products and the development of modular systems for allowing new forms of organization of production, including new spatial forms.

(b)   Transformation, also, of the horizontal production structure, of the systems of relationships between activities or "professions". This transformation results on the one hand from new technological interdependencies and convergences, and on the other hand from transformations of demand, forcing the firms to offer ensembles of products (series of similar products, ensembles of complementary products, combinations of goods and services). The production of these complexes of goods and systems implies the combination of multiple techniques and know-how.

Still, the organizational dynamics of contemporary C&C firms tend to avoid the reality of technological development and demand behaviour. Organizational dynamics also follow the dynamics inspired by the pursuit of functional integration of all business functions, including those beyond the production chain. Moreover, they are affected by movements in the stock market, business history and culture, as well as regulation of the business environment. In the next chapter, we attempt a better understanding of organizational dynamics "beyond manufacturing".

CHAPTER THREE
Accumulation and organization in computer and communications industries: a regulationist approach

Frank Moulaert & Erik Swyngedouw

The objective of this chapter is to develop a coherent perspective upon the processes of regulation and organization of industrial firms in general and C&C firms in particular. For this purpose, we use the theory of regulation and its methods of analyses as developed by Aglietta (1976), Boyer (1986), Leborgne & Lipietz (1988) and others. This approach provides a useful general means of analysing sectoral and industrial dynamics, as their organization and development are affected by changes in the regulatory stance of contemporary states and societies.

In the first section, we summarize the theory of regulation and point out its analytical value. We especially stress the importance of the notion of institutional forms as an intermediate concept, allowing links to be made between, on the one hand, the general logic of accumulation and regulation under capitalism and, on the other, concrete sectoral organizational dynamics and corporate strategies. In the second section, we apply regulation theory to sectoral analysis. First, the regulationist approach is used to show how closely intertwined are the dynamics of institutional reform and the sectoral composition of social production in the changing regime of accumulation. Then, an attempt is made to reproduce the accumulation–regulation dynamics at the sectoral level. This is done from an historical perspective. The third section refines the regulationist approach to the study of sectoral dynamics in the contemporary economy and applies it to the C&C sector. Changes in consumption norms (more customized and informed consumer demand), functional integration and flexible forms of economic cooperation and spatial networking, lead to the conclusion that the flexible production system is probably as flexible (and therefore largely unpredictable) in its business-organization dynamics as it is at the level of the production process itself. However, it should be stressed that these conclusions are based on information for large and very large leading enterprises. Therefore, they should be interpreted with care.

Regulation theory: its value for sectoral analysis

Regulation theory attempts to link the structural dynamics of capitalist society to concrete institutional dynamics and corresponding agency activity in the social and economic lives of individuals. For this reason, explanatory categories in regulation theory are shown to be applicable to the analysis of sectoral dynamics.

The innovative character of regulation theory

Following Boyer (1986), Leborgne & Lipietz (1988), Moulaert, Swyngedouw & Wilson (1988) and Moulaert & Swyngedouw (1989), the basic strength of the French regulation approach is the way in which it manages to model society by integrating the logic of capital on the one hand, and the individual and social behaviour of agents in that society on the other hand. This integration is achieved through the application of a set of intermediate abstract categories, i.e. regime of accumulation and mode of regulation, both embodied in historically specific, but dynamic *institutional forms*.

Institutional forms are the concrete expression, the embodiment of the structure of society in a given historical period. The institutional forms of society, therefore, cannot be *a priori* theorized but must be historically constructed. They are significantly conditioned by the functioning of society while, at the same time, they facilitate its reproduction in a dynamic way. But institutional forms are equally well the result of individual and collective agency acting within and upon the historically produced social configuration. Therefore, the introduction of the notion of institutional forms endows the theory of regulation with the analytical capacity to study the reproduction of the socio-economy in its actual institutions. These institutions include economic institutions such as firms and industrial sectors.

40

The model of society according to regulation theory

According to the regulation school, the analysis of society and its institutions should be made at three, interactive levels: (a) the level of the mode of production; (b) the level of the economic system or the regime of accumulation; and (c) the level of institutional forms, i.e. the mode of regulation. While the analysis of the dynamics of society is situated on the first level, the analysis of the actual forms and mechanisms of socio-economic reproduction of society is situated on the second and the third levels. Together, and considered interactively, the mode of regulation and the regime of accumulation constitute a development model.

The first level, i.e. the *mode of production*, is concerned with the deep structures of social organization. In capitalist society, these structures revolve around the process of value production through the circulation of capital, a process which is predicated upon the private accumulation of capital by means of the exploitation of labour. The contradictory dynamics of this mode of development contain simultaneously the possibility, or often the inevitability, of crisis, as well as the plausibility of some medium-term dynamic equilibrium. The theoretical analysis of the mode of production as a complex real and conceptual category is essentially a theory of crisis. The theoretical expression of crisis and its dynamics necessitates the possibility of some forms of dynamic equilibria. In other words, given the crisis-ridden nature of the capitalist mode of production, the key problem is to understand the mechanisms maintaining an unstable dynamic equilibrium which preserves the deep structure of society.

Two elements are important in this respect. First, history has to be simultaneously maintained and transformed: maintained in the sense that the deep structures of society have to be reproduced; transformed from the moment the reproduction of the deep structure of society is threatened. This entails two things. On the one hand, the process of capital circulation in its elementary form should continue, while its forms are continuously revolutionized, striving to overcome the threat of transformation of the deep structures of society. On the other hand, the information necessary to maintain this process has to be conveyed throughout the system, from the individual agent to the deep structures and from the structures back to the agent. These two processes are precisely embodied in the notions of *regime of accumulation* and *mode of regulation* respectively. The first assures the dynamic reproduction of the circulation of capital; the second harnesses, codifies and informs the accumulation regime. Indeed, the institutional forms assure the conveyance of information (conceptualized as bits of systems-maintenance capability) throughout the various layers of society.

41

As Boyer argues, a *regime of accumulation* comprises:

"The ensemble of regularities that assure a general and relatively coherent progression of the accumulation process. This coherent whole permits the absorption or temporary delay of the distortions and disequilibria that are born out of the accumulation process itself." (Boyer, 1986, p.46) [our translation].

These regularities are of a social and economic nature and they concern:

(a)   the type of development of the organization of production and the relationships of the production agents towards the means of production; these can be described in terms of a technological paradigm and the mode of organization of the labour process (cf. Dosi 1988, Leborgne & Lipietz 1988);

(b)   the time horizon for returns on capital investment on the basis of which management principles can be determined;

(c)   the distribution of revenues from production to different social classes and groups;

(d)   a certain composition of social demand guaranteeing investment in further production capacity;

(e)   a mode of articulation with older or less advanced regimes of accumulation, in case these occupy a determining position in the dynamics of the economy in question. The latter is very important for understanding sectoral and spatial development.

The *mode of regulation* can be defined as the "complex of procedures and behaviours, individual and collective, possessing the triple quality of: (a) reproducing fundamental social relationships by interaction and co-operation between historically determined institutional forms; (b) supporting and guiding the developing accumulation regime; (c) guaranteeing dynamic compatibility between sets of decentralized decisions, without requiring economic agents constantly to be told the principles of self-adaptation appropriate to the system" (Boyer 1986 [our translation]). In other words: the mode of regulation communicates information throughout society via its institutional forms from the micro-level to the macro-organizational structure and back.

The regularities associated with, and embedded, in the institutional form are explained not in terms of a general logic of capital (although their functioning can be derived from the latter), but by analysing that general logic through the forms adopted by the economic structure during successive historical periods of its development. These forms include, among others, the changing forms of competition, wage–labour relationships, regulation of the financial and monetary system, and the structuring of the international economy. More precisely, the identification of accumulation regimes is not only predicated upon the analysis of the economic structure, but also of institutional forms: (a) forms of organization of capital and money markets; (b) forms of wage–labour relationships (historically characterized by five dimensions: type of means of production, forms of technical and social division of labour, forms of mobilization of workers and their identification with the enterprise, determinants of direct and indirect wages, and consumption and life-style norms); (c) forms of competition; (d) state and supra-state regulation.

The precarious coexistence of a regime of accumulation and mode of regulation is under continuous pressure, not only because of the contradictory nature of capitalist development *sui generis*, but also because of institutional dynamics themselves. Under conditions of accelerated build-up of crisis, institutional forms become contradictory themselves in terms of their ability to maintain the deep structure of society and to convey the system's information through the various scales and layers of societal organization.

In short, the regulationist approach theorizes, first, the social and economic forms through which previous contradictions in economic and spatial development (and the methods of resolving these) become problematic in themselves and provoke a developmental crisis; and secondly, it defines the development of new socio-economic forms resulting from the crisis process and the actions taken by (groups of) social agents. Embedded in this approach, then, are possibly different forms of crisis:

(a)   short "conjunctural" crises requiring only minor adjustments, such as incremental technological changes, expanding (spatial) divisions of labor, institutional adjustments;

(b)   structural crises (or crises of a particular mode of development) leading to qualitative changes in the organization of the accumulation process;

(c)   crises resulting from the fundamental contradiction of the capitalist mode of production itself.

The regularities and forms in Table 3.1 correspond basically to a reading of social dynamics at the macro-level of society. Such a reading lays a

Table 3.1 Regime of accumulation and mode of regulation: regularities in economic structure and institutional forms

Economic structure	Institutional forms
(a) relationships between forces and relations of production: technological paradigm; industrial relationships	(a) forms of organization of wage–labour relationship
(b) a certain type of sector and market organization: market structure, forms of competition, inter-/intra-sectoral relationships	(b) forms of organization of competitive relationships
(c) a certain type of distribution of produced value	(c) forms of monetary and financial regulation
(d) a certain composition of social demand (consumption norm)	(d) state regulation
(e) a certain social and spatial division of labour	(e) insertion in the international system

basis to help us to understand the meaning of social dynamics and analytical categories at the sectoral level, to develop categories that facilitate the analysis of organizational strategies at the firm- and sector level, and to apply these to specific sectors.

Regulation theory and sectoral analysis

In the first of two subsections, the regulation approach shows how the dynamics of institutional reform are closely intertwined with the sectoral composition of social production in the changing regime of accumulation. In the second subsection, an attempt is made to reproduce the accumulation–regulation dynamics at the sectoral level. This is done from an historical perspective, by looking at the technological and organizational trajectories of leading sectors in successive accumulation regimes.

The dynamics of institutional reform and sectoral shifts
Sectoral shifts and changes in institutional forms do not take place in a vacuum. They are conditioned by, and they respond to, the failure of existing forms to deal with the contradictions of capitalist development. Take, for example, the dynamics of Fordist accumulation and regulation: how should we understand the response of capitalist accumulation to the Fordist institutional forms? What has been the sectoral articulation of that response?

44

Without going into detail about the characteristics of the Fordist institutional forms, it is important to summarize two key elements. First, the regulation of the production system, market structure, and wage–labour relationship was institutionalized on a national basis. It was the nation-state which codified the mechanisms of social reproduction through nationally organized Keynesian demand management, institutionalization of class relationships, and sectoral policy. This resulted in a highly divergent institutional structure among nations. Secondly, the international institutional forms were weakly developed and they emphasized trade and monetary issues. Apart from a rare attempt to regulate production and social reproduction on an international level (the Marshall Plan, for example), it was mainly the financial system which was harnessed in a series of regulatory mechanisms. Some of these remained fairly weak (see in particular the failure of the IMF and the World Bank to keep the credit system under control); others proved to be of paramount importance, i.e. the regulation of the monetary system (Moulaert & Vandenbroucke 1983).

The national mechanisms of production and market regulation, demand management and wage–labour regulation often resulted in geographically widely divergent conditions of social development and economic performance. In contrast, the establishment of the international monetary and financial system was meant to contribute not only to general stabilization, but also to harmonization of trade and finance at the world level. The interplay of these national and international regulatory dynamics produced a basically fragile harmony between accumulation and regulation conditions. This fragility had two important implications. First, financial capital – capital by means of financial operations and speculation – was limited in its powers to accumulate. Moreover, variations and fluctuations in economic performance between countries were not reflected in changing relative strengths of the countries' currencies. The dollar was the *de facto* world currency, guaranteeing US supremacy in the world market. The relative strength of the other Western economies was measured by their access to world finance, and consequently by the strength of their productive apparatus. This in itself explains the feverish drive for productivity increases. The second implication of this fragile equilibrium was the requirement for stability of the economic system, which put a premium on accumulation by means of industrial capital rather than by expansion of financial speculation.

Put simply, the combination of these two institutional forms – i.e. national demand management and sets of international financial stability – promoted capital accumulation by means of commodity production. Hence

45

the rapid expansion and growth of manufacturing industry during the Fordist era.

Moreover, the combined dynamics of these institutional forms transformed the world from a fragmented production space of national territories competing for each other's markets, into a system of fragmented production and consumption spaces competing for "world" capital. In other words, the political and economic command over space which coincided with the borders of the state in the pre-Fordist period broke down as a result of the institutional forms coming into place after the Second World War. It was replaced by a global space of production, but with a fragmented nation-based regulation of market structures and wage–labour relationships. In sum, the Bretton Woods agreement of 1944 not only promoted commodity production and exchange but predicated trade expansion upon the penetration of highly uneven, but internationally linked, spaces which made up the capitalist world economy. The resulting international spatial division of labour challenged the competitive position of the core countries. It also affected their ability to regulate markets and management–labour relationships through national demand-management policies. In particular, the USA reacted by withdrawing from the Bretton Woods agreement that had largely facilitated the expansion of the system. The ensuing breakdown of the international monetary order encouraged the rise of internationally organized commodity production by means of a spatial division of labour and production for an international market. International monetary instability (resulting from the change in the institutional framework) induced a rapid sectoral shift from manufacturing, requiring long-term fixed investments, to services and, in particular, financial speculation. At the same time, this very instability required an extremely flexible and fast response in many activities, and an increasing pressure on the international monetary order towards a more deregulated structure. This transition alone is an important element in understanding the rapidly growing demand for information, information networks, -processing and -advice.

Intrasectoral recomposition, then, is equally associated with this transformation of the financial system. What we have been confronted with since the early 1970s is a highly fragmented but interlinked geographical system, operating according to rules different from those under Fordism. Each spatial fragment constitutes both a market and an internally homogeneous monetary system, either nationally organized (yen, pound), or internationally (ECU to some extent, US dollar). All fragments operate in an asymmetrically organized monetary system of monetary exchange linking up the fragments of space. Manufacturing industry, in particular its leading sectors, is confronted with a major contradiction. On the one hand, the long-term survival of these sectors is predicated upon the ability of their firms to penetrate and capture new markets; on the other hand, the production process is necessarily fixed in space. Under Fordism, cost control and market expansion via direct foreign investment proved to be the safest route; a strategy now becoming extremely risky and with unpredictable medium-term outcomes.

The intrasectoral recomposition of the C&C sector illustrates the way in which leading-edge sectors reorganized in the face of this institutional form. In particular, the restructuring of the interplay between local and global factors is of paramount importance here. The internationalization of production and the penetration of new geographical markets is no longer pursued solely through new direct investment, but also by the setting up of an international network of strategic alliances and well selected acquisitions within and between monetary zones. They facilitate market penetration without running the risks associated with the long-term fixation of capital through direct foreign investment in a particular place. Less direct foreign investment, and much more indirect bilateral market penetration, is the way in which the spatial expansion of the system is organized. Nevertheless, the production process remains necessarily fixed in space. The minimization of risks associated with the attachment of the production process to space is pursued by strategic rethinking of internal and external business organization. This can include externalization of non-key functions of material production, or spatial spreading of complete production networks in a series of places. At the same time, the internal or internalized functions and processes are fine-tuned to accommodate rapid changes in national and international regulatory conditions, and in technological requirements. The model of the "hollow corporation" (intensive in terms of knowledge and information control, extensive in terms of low value-added material or services production) is the penultimate result of this re-definition of the interplay between local and global factors which takes place in the turmoil of a changing international regulatory structure. However, this model remains an ideal which, while informing strategies, is nevertheless not realized uniformly in practice.

Regulation theory and sectoral dynamics: an historical classification
Historically, successive regimes of accumulation have been characterized by the systemic forms taken by technico-organizational structures or technological trajectories of their leading sectors (Dosi 1988, Freeman &

Perez 1988). Each trajectory (or developmental profile) varies with the dominant paradigm of corporate organization practised within the firm. Moreover, it varies in relation to products, markets, competition, production processes and the geographical location and organization of the firm. Table 3.2 illustrates these historically successive forms of technological and organizational trajectories and their corresponding characteristics in terms of the organization of the labour process. The table poses the question of, on the one hand, synchrony and diachrony in theorizing both the transition from one form to another and, on the other, the possibilities of institutional macro-stabilization. Although we can easily identify historical periods of relative dominance of one sectoral technological trajectory over the other, each successive moment still contains important fragments of the previous forms, adding to the overall complexity of the system.

Therefore, each regime of accumulation is characterized by the dominance of the latest trajectory over the previous. But, at the same time, the other forms persist, in a more or less pronounced yet transformed way. The relative importance of a new trajectory depends, among other factors, on its adaptability to new conditions, the relative strength of the dominant trajectory, and the integrative capacity of institutions. Indeed, institutions unite and link the various trajectories, which enables the system as a whole to function.

This shows again the importance of institutions and of the mode of regulation, in defining the mode of development. Indeed, however important technico-organizational trajectories themselves might be in terms of sustaining the accumulation process, they are insufficient to explain the unity of the system as a whole in which various trajectories co-exist. In fact, it is the dynamic interaction of institutions with the dynamics of technico-organizational change which establishes the dominance of one system over the other, as well as regulates the interaction between the various subsystems.

For example – and quite important in understanding the transformation of the C&C sector – the Fordism versus flexibility debate transcends the replacement of one system by the other. Instead it expresses theoretically the "transition" from one form of dominance to another. However, the resulting change in the mode of development depends on the way in which the institutional forms are re-created in particular places at particular moments in time. This process of re-creation of the institutional forms is in turn predicated upon the relative strength of the various dominating and

Table 3.2 Historical modes of social and technical organization of the production process.

Nature of labour process	Division of labour	Means of production	Technical relations of production	Labour relations of production	Type of enterprise
Individual	None (artisan)	Simple tools No task diversification	Qualified labour Unified LF & ML Unified LF & PP	Self-employed	Artisan
Co-operative	None	Simple tools No task diversification	Qualified labour Unified LF & ML Separate LF & PP	Class-based Stratified authority	Simple co-operation
Co-operative	Technical (textile, 19th century)	Simple tools Simple task diversification	Semi-qualified labour Separate LF & PP	Class-based Stratified authority & skill	Manufacture
Co-operative	Technical (steel, coal)	Machines Complex task diversification	Qualified labour Separate LF & ML Separate LF & PP	Class-based Stratified authority & skill Separate mental & manual labour	"Modern" industry Taylorism
Functional separation	Technical Spatial (automobiles)	Complex machines Complex task diversification Task demarcation	Qualified labour Separate LF & ML Separate LF & PP	Regulated stratified Class-based Institutionalized labour-force Segmentation Bureaucratization	Fordism
Functional and organizational separation	Technical Spatial (informatics)	Complex automated machines Knowledge & science Systems integration	Qualified labour Unified PP Innovation process	Multi-stratified Flexible & fragmented labour-force Organization & regulation	Flexible

Abbreviations: LF for labour force, ML for means of labour, PP for production process.

subordinate social classes and the mixture of alliances among them. The relative success of a particular combination of a technical organizational trajectory with institutional forms is, then, checked *a posteriori* by its success in reproducing the deep structure of the system as a whole (Leborgne & Lipietz 1988).

The timescales of regimes of accumulation, and corresponding modes of regulation, as well as the classification of sectoral trajectories, remain extremely difficult, in particular in periods of transformation during which the relative success of one mode depends on the strength of the other. The classification is made even more difficult by the persisting co-existence of various trajectories. In the end, it is the nature of the institutional forms which imposes the mode's unity on the other co-existing forms.

Therefore, the definition and analysis of a sector largely transcend the rather narrow technical-organizational principles, but should include the particular regulatory forms dominant during the period and in the areas in which the sector develops. It is these forms that impose the mode of development of the system as a whole. For example, while it is clear that artisans and small-scale workshops based on a simple division of labour did not disappear during Fordism, Fordist institutional forms dramatically affected their further development. The regulatory forms affected their operations as much as any other enterprise, while welfare consumerism altered their market conditions significantly. In fact, one could argue that it was precisely the regulation of the wage–labour relationship which simultaneously undermined the competitive position of some of the old organizational forms and created a particular recomposition of the consumption norm, facilitating the rapid expansion of, for example, artisan-based flexible production.

A problem is posed not only by the neglect of the broader institutional forms in the timescale of regimes of accumulation and their typical sectoral trajectories. The notion of technico-organizational structure by itself has become insufficient to cover the organizational complexity of the Fordist and post-Fordist firms and sectors. This notion overstresses the importance of the technological system and the manufacturing labour process, without taking into account other entrepreneurial functions such as marketing, distribution, management and administrative functions, logistics, etc. The study of the functional separation of the labour process is too much confined to the production function itself. Speaking ironically, one could say that in the literature there is a dominant tendency to make a Taylorist reading of Fordism and post-Fordism. This analytical anomaly transforms the functional and organizational dynamics of transnational, multi-functional enterprises into a division between intellectual and manual labour, and between different manufacturing functions and tasks. This Taylorist bias in different social science disciplines and paradigms (also partly in regulation theory itself) leads to a misunderstanding of regulatory mechanisms within sectors and firms. For example, it reduces to production-line problems the interpretation of organizational contradictions within the Fordist production system, and it strongly underestimates problems related to business bureauracy (e.g. over-centralization), interaction between business functions, markets, and so on.

Regulation dynamics in a contemporary sector

In this section, an analysis of accumulation–regulation dynamics is developed for economic sectors typical of the flexible (or often called flexible) accumulation regime. Consequences for the analysis of the C&C sector are pointed out. On the basis of the previous discussion, we can conclude that particular regulatory dynamics at the sector level can be identified by combining: (a) technical-organizational structures, a notion to include broader functional dynamics of firm and inter-firm organization; (b) regulatory forms typical of a period of economic development (with a regime of accumulation and regulation) and in which the sector in question plays a leading part in the accumulation process.

The interactive dynamics of economic structure and institutional forms at the level of a manufacturing sector are schematically represented as in Table 3.3. Rather than putting all elements in this table into detailed analytical perspective, let us look at the structural developments typical of contemporary sectoral dynamics.

The consumption norm
Let us look first at an element not explicit in Table 3.3, i.e. the changing consumption norm. A consumption norm is a summary label referring to the different dimensions of market demand in a society, a community, etc. Just like sectors or products, forms of regulation typical of a regime of accumulation do not disappear when a new stage of development emerges, but are rather reshaped or integrated in the new stage. Consumption norms in a new economic era are profoundly affected by those existing before. In this respect, if we accept that we live in, or are heading towards, a post-Fordist era, the post-Fordist consumption norm is still to a large extent a mass-consumption norm. Final demand remains determined by the mass-

Table 3.3 Business functions and institutional forms for leading innovation sectors in the contemporary economy

Entrepreneurial functions	Technological trajectory/content	Work process	Regulatory forms
Management hierarchy/control & co-ordination Financial operations	Information & communications technology (ICT)	High professional work; combined job hierarchy and project-wise work organization	National & international agreements, laws & regulation re mobility of capital & labour/factor income National & international legal forms of enterprises Co-ordination centres, economic free zones
Research & development	Sector-inherent technology: – electronics – telecom – bioengineering – etc. products process technology	Interactive project-based work organization	R&D policy technology (patents & licences, standards & norms, etc.) National & international legal forms of enterprises
Manufacturing engineering	ICT product process	Highly motivated (robots/conveyor transports/CNC) Polyvalent Highly skilled "core" workforce; small peripheral & flexible labour force	Labour & social laws
Assembly	ICT product process	Conveyor transport partly automated, partly manual assembly Highly flexible, medium-skilled labour force	Regulation of flexible labour market
Marketing	Logistics (transportation & ICT) ICT	Skilled commercials, highly flexible work organization	Regulation of flexible labour market Regulation of markets: standard norms/liberalization of trade
Operational administration	ICT burotics	Highly skilled clerical workforce Administrative robotization	Labour & social laws
Personal & maintenance services	Low technology	Low-skilled, highly flexible labour force	Labour & social laws
Professional	ICT burotics	Highly skilled professional labour force	Labour & social laws Regulation: mobility of capital & labour

production system, the large advertising campaigns, the materialist way of satisfying elementary needs, the wage–labour relationship, be it now more deregulated than under Fordism (sectorally and territorially decentralized, for example).

But at the same time, the consumption norm has changed. The satisfaction of individual needs has become more than an illusion or simply the result of publicity campaigns. Varieties of goods and services, more in correspondence with individualized demand (which is not to say individualized real needs), are the combined results of market strategies, production possibilities and consumer behaviour.

It is not possible to detail these developments in the present context. Suffice it to say that consumers have become more conscious, not least as a consequence of the marketing and advertising strategies and the consumer response to it. Consumers have become much more rational. As we saw in the previous chapter, the diversification of products has become an active marketing strategy, and the flexible production system is able to feed that strategy. If consumer markets have become capricious, Fordist production strategy and market regulation are largely responsible.

The interaction between customized, varied supply and the flexible production system is a very important indicator of the new sectoral dynamics. As pointed out in Chapter 2, standardization of components and flexible reprogramming of processes allow for the use of basically identical technical products to satisfy different types of demand. The technical base products satisfying intermediate demand in the C&C sector, for example, are not fundamentally different from those needed to satisfy final demand. Differences will lie in applications software, as well as the mechanical and communication (optics, voice) extensions of the equipment. Given the high level of intermediate demand in such new sectors as C&C, this observation is quite important.

Another aspect of the changed consumption norm, which might even be more typical of the C&C sector, is the "global solution". Clients are eager to receive a global solution for a set of related needs, leading firms to develop a corresponding "global solution" strategy. Such a strategy aims to provide a customized package of equipment and services which satisfies the particular and complex needs of the client with respect to C&C. The development of multi-purpose equipment and services, flexibly adaptable to the needs of the clients, as well as to the new organizational style of the suppliers, render such strategy viable. The global solution strategy (see Ch. 8) allows the client to challenge the C&C supply system to its deepest organizational roots.

53

Business functions: integration, disintegration

From our research we have learned that it is impossible to characterize the emerging regime of accumulation and its leading sectors in terms of a simple, finite list of functional and institutional forms regulating all business tasks. The new sectoral dynamics are not as dominated by vertical disintegration as much of the literature on industrial reorganization suggests. This fallacy has been pointed out by Martinelli & Schoenberger (1990): oligopoly is alive and well. The technical delimitation or disintegration of business functions (see Table 3.3), the curtailment of the corporate business management system and firm-wide labour regulation, do not mean that the social relations of production (the control sphere) in which functions operate are fragmented, and that under post-Fordism the economy would proceed to a more decentralized market structure. Nor do these managerial reorganizations lead to the dominance of new forms of economic control by capital.

Part of the confusion over this issue arises out of terminological ambiguity. This ambiguity is complex and, to keep the situation simple, we must use terminology that allows a distinction between reorganizations referring to property relationships and those concerning business functions and their co-ordination. *Horizontal integration* between firms or groups may mean that an acquisition has taken place, and that a previously independent firm belongs to the same, or a parallel, production or service network as the acquiring firm or group. Therefore, according to this meaning of the term "integration" (which is the meaning commonly used in economics) a firm is integrated into a firm or a group when it is purchased by another. *Functional integration* can occur when different business functions are better integrated with each other, as when there is better co-ordination of material and information flows between them, interactive management, etc. This does not necessarily mean that functions become part and parcel of the same business unit or the same firm or group; it does mean that the functions are better matched with each other in view of the overall business objectives of the controlling firm or group.

By looking at the interaction between functional (dis)integration and new modes of control, competition and co-operation, the growing complexity of the processes and methods of control of business activities may be perceived. Functional integration does not imply acquisition or merger. Co-ordination may happen through new internal control systems, replacing traditional business hierarchies, or through co-operative agreements adopting different forms of economic alliances (joint ventures, reciprocal licences, subcontracts [see Ch. 4]).

In general, most of the firms we have looked at tend towards more functional integration: different technical functions, business and/or product lines, and client sectors are better integrated with each other, making them equipped for active market strategies and to control the cost process of business operations better.

In order to be competitive in *client sectors*, to respond faster to client needs, market considerations – ranging from global solution strategies for institutional clients to diversified product supplies on mass markets – are integrated into the different functions of the business lines. Marketing considerations feed systematically into R&D, design and production prototyping. Quality concerns demand a better integration of R&D and manufacturing engineering, as well as systemic quality control in the business flows. The rationales of forward and backward linkages are absorbed in the management of different functions as well as in their co-ordination.

The improved integration of functions and business lines with different client sectors is not only rooted in marketing considerations, but is also inspired by concerns of *technical productivity*: economies of time, generalization of "just-in-time" (JIT) principles from the core production process itself to other entrepreneurial functions and among business lines (Swyngedouw 1987a).

In this respect, the notion of the enlarged or global factory implies the timely delivery of high-quality supplies from one stage in the functional organization of the enterprise to another, as well as among business lines. Of course, functional integration is a matter of not only logistics but also co-ordination, regulation through market mechanisms, agreements, or simple customer-power. In any of these cases, guarantees are pursued with respect to quality, delivery times and prices of received supplies or smooth forward outflow towards users.

The integrated division of labour, whereby different functions can be assimilated in relatively small operational units, has no *a priori* bias towards more or less vertical or horizontal integration in the sense economic science uses it. Still, empirical reality shows that strategies for mergers and takeovers generate innovative dynamism in financial concentration strategies, and also that continuing financial concentration does not contradict other forms of regulating functional integration, through business lines, client sectors and territorial markets. Regulating functional integration in a high financial-risk environment might well encourage the combination of different forms to regulate interaction between business units: for example, combining market mechanisms with non-market

55

mechanisms, such as agreements and control forms of a different nature, outside and inside the firm (see Ch. 4).

Inside the firm, the application of principles of functional integration, such as decentralized purchasing responsibility or a "just-in-time" policy (including quality and timely delivery norms) is meant to solve typical problems of Fordist management schemes (see Janssens 1985). This, of course, involves decentralizing the decision-making system to the responsible business units. Between these business units, different forms of co-operation are elaborated, often based on internal quality norms and transfer-pricing systems.

Between firms, material and information flows are regulated in very different ways. In the market sphere, there might be agreements for the piecemeal purchase of specific supplies. Still, for basic components we find that either wholesale acquisition or strictly regulated alliances are the preferred strategies to assure supplies. The higher the value-added and technology content of the supplies, and the larger the supplier *vis-à-vis* the customer, the more balanced the power relations between supplier and customer. On the one hand, distribution is completely externalized, with strictly controlled licenses for the distributors; on the other hand, the distribution network is fully acquired. Joint ventures and consortia are the preferred institutional forms for joint research projects and for setting norms for technological systems (such as architectures for information and communication systems).

Strict agreements on quality norms and timely delivery are present in most of the forms regulating intra- and interfirm material and information flows. They are crucial instruments in making the functionally integrated business system work. Such agreements can work fully only if logistics, information and communication systems are in place and organized according to CIM principles.

Business culture and organizational strategy

Organizational dynamics are ruled not only by logic based on technological trajectories, market dynamics and functional organization or reorganization; business culture also plays an important part. Enterprises maintaining their original technology and engineering-based tradition tend to develop their worldwide marketing and distribution system through alliances and long-term agreements with specialist distribution networks: they prefer to stick to their core business. In contrast, firms which developed a strong marketing and distribution strategy from the onset usually develop their own distribution function, as an integral part of the enhanced factory system.

As far as the externalization or internalization of business services are concerned, a multitude of factors intervene: availability of skills, potential economies of scale and scope, proximity of the service content to the main business activities, etc. (Martinelli 1988). However, business culture plays an important part here too: differences in the use of external professional service suppliers are determined by corporate traditions ranging from strong self-reliance ("nobody knows our business better than we do") to perhaps exaggerated dependence on outside specialists (ERMES 1988).

A very important element for C&C service firms (and one which will also probably become more significant for other C&C firms) is the cultural content of the products themselves. In general, among different types of C&C firms, C&C service suppliers have moved furthest in offering customized global solutions to their clients. Part of this expertise is the specific cultural content of their products: their consultancy and service style, their methodology, and their way of interacting with the clients (Moulaert et al. 1990). This has important consequences for the internal and external organizational dynamics of firms. Partners or acquisitions should fit the business culture of the parent; the development of new functions spins off from existing ones, and in the same working and marketing style.

With the development of integrated products and global solutions, and the tendency to service diversification, the cultural content of products and working methods will become important to other types of C&C suppliers.

Organizational dynamics and labour
Organizational restructuring of business functions has major consequences for labour, in terms of levels of employment, skill requirements, mobility and training. The reorganization of manufacturing production (automation, JIT, flexible specialization), and the automation of clerical and administrative work, has eliminated a large part of semi-skilled and unskilled work. The greater emphasis on the integration and co-ordination of functions requires interface capacity, organizational and management knowledge, and professional service logistics in general. This leads to external as well as internal service development, with a corresponding relative increase in employment in R&D and distribution functions, often at the expense of manufacturing employment. In any case, many new professional service jobs inside and outside C&C firms have been created

(Noyelle 1989, Gadrey 1990). There are always new problems to face. The number of lower-level jobs, especially in personal services, catering, cleaning, etc., continues to grow. Since they are meant to facilitate the functioning of firms, employees and their families, they too can be considered as belonging to the "enlarged factory system".

The multiplication of matrix management and project-development strategies, and the pursuit of global solutions tailored to client needs, require increased interactions within business systems, including exchange of information (communication), co-operation, and multiple skilling. The generalization of these principles leads to job enrichment, but equally to increased stress.

The introduction of new technologies and modes of organization, communication requirements and multiple skilling all require continuous (re)training. Training is an important issue to most of the firms we interviewed. It is a key element of the human-resources management most of them have put in place. Training is organized on the job, and in co-operation with sister plants or offices, and also through closer communication with education and training systems.

Spatial organization

Geographically concentrated production and business systems are not the dominant spatial form of the "enlarged factory system". Functional integration may mean – and we have seen it in many cases – that R&D activities are located next to major manufacturing facilities, and that specialized service suppliers might share the same sites. But it does not mean that suppliers tend to locate on customer sites. On the contrary, even if the tendency to establish a worldwide supply system for basic components by one or a few multinational suppliers has weakened, it certainly still holds at the continental or subcontinental level, and many factories see the local or regional part of the value of their supplies fall spectacularly.

Geographical globalization of business systems affects not only supply systems. R&D units may be located more in conjunction with plants, but local R&D activities are part of a supra-regional, often worldwide, manufacturing network co-ordinated by a few core research centres. Factories often have "sister plants" on other contintents, and they operate in close connection with them, as well as with their own suppliers and distributors. Worldwide distribution systems are an important issue. Yet the organization of these systems differs substantially between firms and sectors. The concentration in distribution grows rapidly, through both the acquisition of specialized distribution networks by main C&C firms and the merger movement in the C&C distribution sector itself. Marketing and sales practices are internationally harmonized, all with at least a minimum of respect for national and regional culture and for user needs.

C&C industries and services do not offer better opportunities for regional development than the traditional Fordist branch plant economy. Certainly, some old manufacturing regions have been successful in attracting high-technology manufacturing and R&D activities (Amin & Goddard 1986). But the development potential from remote areas or third-tier urban centres (via emerging C&C industries and related services) is really limited. And, for those regions which have been studied in depth, a reproduction of regionally unequal job opportunities is often the result of C&C investment strategies. (See Martinelli 1986 on professional services in Italy, and Swyngedouw & Andersen 1987 on C&C industries in France).

Back to social regulation

The dynamics of C&C industry and services are a major challenge to regulation internationally, at the level of nation-states, and also for regional and local governments. The question is, then, to what extent does deregulatory behaviour at the firm level (wage–labour flexibility, administrative decentralization, functional integration, etc.) stimulate changing regulatory behaviour at the different state levels.

The answer to this question is complex. State regulatory behaviour is obviously not a mere public affirmation of individual firm behaviour and deregulatory requests. The rôle of the state transforms, but does not simply act as the instrument of, corporate interests. Several functions of the state have changed, and levels of decision-making have shifted. With respect to market regulation, the international dimension has become more important. Supra-national organizations, such as EEC and OECD, often in co-operation with international professional organizations, play an important part in the liberalization of trade, the establishment of uniform technical and quality norms, and the simplification and harmonization of administrative practices. In this way, national states have delegated some of their power to supra-national decision-making bodies, enabling them to promote the globalization of markets.

In many countries, industrial policy (with the exception of fiscal measures) has devolved to regional and local authorities. And even if these authorities do not fully possess the financial means to carry out such a policy fully (e.g. for the acquisition of large-scale equipment for telecommunications and transportation systems) and should therefore find recourse to the national states, private capital and international organizations, they have often become partners in and orchestrators of local and regional investment projects. Such projects are regarded as contributions to regional development strategies. This transformation of the national Keynesian subsidizing state into the regional entrepreneurial partner state fits the tendencies of capital to develop partnerships and networks at decentralized functional levels, not inhibited by heavy bureaucracy (Moulaert et al. 1988). Still, national states continue to play a major rôle in sectoral policy through fiscal deregulation and, perhaps even more so, through their position as, for the time being, monopoly purchasers in many public telecommunications and transportation infrastructure markets. This position allows them to foster C&C "national champions", a policy whose fragility should become clear in the next decade.

Wage–labour regulation remains largely with the national authorities. But the deregulation tendency is strong, and social protection less assured than under Fordist social policy. Much credit is given to capital and private investment strategies to assure employment and income for all. But since C&C industry and services do not guarantee equal inter-regional development opportunities, this *laissez-faire, laissez-aller* wage–labour policy will not suffice as an engine of prosperity for regions falling beyond the core geography of the C&C sector.

The globalization of markets, the spatial integration of production systems, leaving substantial autonomy to lower levels of decision-making in business administration, and thus creating opportunities for local development strategies, stimulate reflections on the prospects for supra-national partnership and economic development policies. These mainly theoretical points will be explored in subsequent chapters of this book.

CHAPTER FOUR
Globalization and its management in computing and communications

Philip Cooke & Peter Wells

Globalization is probably the most powerful force affecting the practices of firms, from the largest to many smaller and medium-sized enterprises in the present period. The term "globalization" is widely used but inadequately specified at present. In a relevant book (Guile & Brooks 1986) it is argued that the two main forces responsible for it are: (a) the substantial relative growth of the European and Asian economies as compared to that of the previously dominant USA; and (b) the growth of transnational corporate activity in both production and distribution as a consequence of the growth of regionalized global markets.

As a consequence of the degree of interpenetration by transnational corporations into the "Triad" markets (Europe, Japan and the USA; Ohmae 1985), there is a degree of re-balancing going on between them. This is not to say that the level of interpenetration is equivalent – the US is the most open, the Japanese the least – so significant levels of sectoral and spatial uneven development remain.

However, it has to be said that this is a rather limited view of the nature and intensity of globalization processes. It says little about the reasons for trade growing between global markets, or whether there is a change in the nature of that interpenetration. Nothing of significance is said about the phenomenon of the new international division of labour, or its counter-tendencies, and inadequate attention is paid to the rôle of governments and supra-national regulatory agencies in the process. The dominant view is that markets are driving globalization, and new products – especially technologies – are driving markets.

In what follows, five key features of globalization will be identified and discussed: markets, finance, location, technology and regulation. Each of these may be seen influencing globalization in distinct though related ways. Thereafter, attention will be drawn to the particular intersection of these elements in the case of the industry branch of most significance in the subject under scrutiny: information and communications technologies. Then, a particular form of globalization which is neither purely market nor hierarchy (Williamson 1985) but rather a complex intermediate form – the strategic alliance – will be picked out for examination. Conclusions will then be drawn from these discussions.

Trade and markets

If we take manufacturing to begin with, there is good evidence of growth in the world market for many key sectors, as Table 4.1 shows. In general it is clear that globalization is resulting in a relative decline in American market power *vis-à-vis* Europe and Asia, but the USA still remains the dominant market. In production terms the US share of the "Triad" fell from 51.6% (1984) to 46% (1986) in four key electronic sectors (Table 4.2). Although Japan increased its share of production, Europe grew most from

Table 4.1 Growth ($ billion) in global manufacturing: leading sectors and UK.

	$ bn 1986	% change 1980–86	% UK 1980	share 1986
Telecom equipment	31	233.2	4.8	5.3
Office equipment	25	230.7	11.5	9.1
Sound recorders, photograph	14	161.5	8.3	6.1
Automatic data processing equipment	35	161.0	13.2	11.1
Transistors, valves	29	85.2	5.5	6.4
Women's outerwear, non-knit	17	79.8	6.3	5.2
Motor vehicles	107	79.4	8.2	6.6
Outerwear knit	15	78.6	7.2	5.5
Engines & motor	12	74.5	19.4	15.4
Measuring & controlling instruments	23	63.6	10.1	9.4

Source: PA Consultants (1987).

Table 4.2 Production and market levels in electronic data processing, office equipment, telecommunications and electronic components, 1983–87.

(us$ million)	1983	1984	1985	1986	1987
Production					
USA	85,693	111,115 (51.6%)	106,622 (48.6%)	105,052 (46%)	NA
Japan	42,524	55,077 (25.6%)	58,553 (26.7%)	63,236 (27.7%)	NA
Europe	NA	48,960 (22.8%)	54,436 (24.8%)	60,124 (26.3%)	66,201
UK	NA	8,443 (3.9%)	10,085 (4.6%)	11,349 (5.0%)	12,482
France	NA	9,403 (4.4%)	9,744 (4.4%)	10,683 (4.7%)	11,784
TOTAL	NA	215,152	219,611	228,412	NA
Markets					
USA	108,062	105,316	108,485 (51%)	142,256 (53.9%)	NA
Japan	36,221	37,566	39,883 (18.8%)	51,432 (19.5%)	NA
Europe	NA	NA	64,026 (30%)	70,445 (26.7%)	76,971
UK	NA	NA	12,453	13,735	14,972
France	NA	NA	10,778	12,943	12,673
TOTAL	NA	NA	212,394	264,133	NA

Source: Mackintosh (1986, 1987).

22.8% (1984) to 26.3% (1986). The position with respect to markets is less clear, since fully comparable figures could be obtained only for 1985 and 1986. 1986 saw US markets grow nearly 30%, but this has proved to be a temporary surge. Subsequent years have seen a return to the very low levels of growth in the total US market shown in the early to mid-1980s.

These trends are underlined in Table 4.3, which shows aggregate market and production change over the longer term, 1975 to 1985. Japan has the highest growth rates, France the lowest, with the USA, UK and the Federal Republic of Germany (FRG) falling in between. Over the 1980s, in the four key sectors shown in Table 4.2, the UK has marginally improved its share of production and markets compared to France. The explanation for these relativities is straightforward. Japanese output is generally both of higher quality and reliability per unit of cost than those of its competitors. On the one hand, Japanese production output has consistently grown faster than Japanese markets; on the other, US production growth has consistently lagged behind the growth of US markets. Consequently the USA has moved from a trade surplus in telecommunications products of $275 million in 1982 to a deficit of $1.7 billion in 1986 (Enderwick 1989). In electronic data processing, the US output in 1984 was $36 billion while the market was $41 billion, a difference of $5 billion. By 1986 the difference had soared to over $15 billion (Mackintosh 1987). The competitive struggle is resulting in Japan taking ever larger shares of global markets, especially in electronics and other metals and engineering industries.

Table 4.3 Aggregate market and production growth (electronics) selected countries 1975–86

(real annual rate 1975–86, %)	Market growth	Production growth
Japan	10.0	10.2
UK	9.5	8.5
USA	9.2	8.5
FR Germany	8.2	7.3
Italy	6.3	6.1
France	5.4	5.2
AVERAGE	8.9	8.5

Source: Financial Times, 27 June 1988, p.5

Finance and financial services

The finance industry is, after electronics, probably the most globalized of all industries in terms of growth, though in levels of trade, services generally lag well behind globalization in manufactured products. The repositioning of the world finance industry has led to concentration of financial power in a few global trading centres such as London, Tokyo and New York, but the growth beneath these of a range of international specialist centres and intra-bloc regional centres. As Noyelle's (1986) figures show, trade and the internationalization of services remains relatively small compared to the position in manufactured goods (Table 4.4).

Enderwick (1989) shows that between 1960 and 1982 invisible trade as a proportion of total world trade increased from 23.4% to 29.4%. However, much if not all of this increase is accounted for by foreign travel. Data on the internationalization of different aspects of financial services within the overall services total show that Japanese banks have internationalized very rapidly in recent years.

Table 4.4 Services in the balance of payments in 1982 (US$ billion).

Exporting	nations	Importing	nations
France	40.6	FR Germany	43.4
USA	38.4	Japan	34.1
FR Germany	29.3	France	33.4
UK	28.5	USA	32.9
Japan	20.4	UK	21.9
Italy	20.3	Saudi Arabia	19.7
Netherlands	16.0	Netherlands	15.0
Belgium/Luxembourg	12.5	Italy	14.5
Spain	11.6	Belgium/Luxembourg	11.0
Singapore	9.5	Canada	10.4

Source: Noyelle (1986).

Thrift & Leyshon (1988) show that in 1976 the top five global banks (by market value) were all European or American. By 1981 only three European banks (all Swiss) appeared in the top ten. By 1986, nine out of the ten world's largest banks were Japanese. Indeed, Thrift & Leyshon argue that since 1984 a new international financial system has developed based on the globalization of financial services. Further, they say (p.60):

"Many markets . . . are now 24 hour or near-24 hour (equities) markets. They sell global products which require global marketing and distribution . . . Quite clearly, the new securitised and globalised markets required enormous injections of information and communications technology . . ."

The rise in world significance for the Japanese banks has been driven by the competitive success of Japanese industry. Equally, the decline of the US banks has been paralleled by a decline in the US share of world trade, from 16% in 1965 to 12% in 1985. There is still a considerable hegemony exerted by UK and US accounting firms (the big five), though these have been concentrating as well as proliferating in terms of geographic location.

Thrift & Leyshon (1988) provide some examples. Between 1975 and 1985 the total number of offices operated by the twenty largest accounting firms increased by 115% to 4,991. The regional global distribution of employment among the major accounting firms is typified by Price Waterhouse (Table 4.5). The main explanation for this is that, following a trend in which producer-services firms pursued manufacturing firms to their global locations, financial markets have become global phenomena with internally generated growth characteristics in their own right.

Table 4.5 World distribution of Price Waterhouse staff, 1984.

Region	Total staff (000s)	%
UK	3.1	11.7
North America	12.1	45.7
Pacific & Far East	4.1	15.5
Europe	2.7	10.2
South America	2.2	8.3
Africa, Middle East, Asia	2.3	8.7
TOTAL	26.5	100.0

Source: Thrift & Leyshon (1989).

Location and foreign direct investment

During the 1970s and early 1980s, the crisis of Fordism in the advanced economies led to a notable export of capital from the developed countries to the less developed countries. This resulted in what came to be called the new international division of labour (Frobel et al. 1980). Investment took the form of the establishment of branch factories in low-wage zones in the Far East, Southern Mediterranean and parts of Latin America. Some countries reached the status of newly industrializing countries partly as a result of the stimulus provided by such foreign direct investment (South Korea, Taiwan, Singapore, Hong Kong). However, it became clear at the time that, although such investment marked a change in direction of a proportion of foreign direct investment, the magnitude of the flow was relatively small. By far the greatest proportion of foreign direct investment was undertaken between the developed economies. This flow became overwhelmingly significant during the 1980s; it surged dramatically thereafter with the weakening of exchange controls in European countries (notably the UK) and the rising value of the Japanese currency.

Much of this outflow of capital has been directed towards acquisitions, as in the case of European investment into the USA, with some similarity regarding the rather weaker reverse flow from the USA into Europe. However, a major feature of foreign direct investment (FDI) in the late 1980s has been the establishment by Japan of offshore production and financial services facilities in Europe and the USA. For example, investment by the Japanese automobile industry into the USA during the 1980s reached $6.745 billion (Mair et al. 1988).

Table 4.6 UK capital flows, 1962–84

INVESTMENT (£million)	Inward	Outward
1962	3,160	8,070
1966	4,280	9,935
1970	6,960	14,400
1975	15,103	25,186
1976	17,086	31,592
1977	20,416	32,370
1978	22,441	37,398
1979	26,438	43,063
1980	31,522	51,944
1981	35,840	70,638
1982	38,566	99,130
1983	45,930	120,934
1984	50,141	156,329

Source: Dunning & Cantwell (1987)

Table 4.6 illustrates the growth in foreign direct investment (FDI) for the case of the UK. The USA is the greatest source and recipient of global FDI. Table 4.7 shows how advanced industrial economies have been the principal recipients of outward US FDI, particularly the UK. In comparison, Japan has proven extremely closed to US FDI. While the figures refer to acquisitions, other research has shown that these account for over 60% of all outward investment (Hamill 1988). Conversely, the UK was the largest acquirer of US firms, accounting for 29% of all US acquisitions (618 out of 2,138) between 1980 and 1986. Not only did the numbers of international acquisitions increased rapidly in the 1980s, the value of such activity grew even further. Hamill (1988), for example, shows that UK investment in the US rose from $229 million in 1976 to $8,014 million in 1986.

Table 4.7 Takeovers by US companies 1980–86, by country.

Country	80	81	82	83	84	85	86	Total
UK	38	27	30	39	45	44	50	273
Canada	15	8	30	35	24	42	54	208
Germany	7	11	12	13	13	15	13	84
France	8	12	8	8	15	16	12	79
Japan	—	—	3	3	7	7	2	22
Others	34	43	38	48	43	51	49	307
TOTAL	102	101	121	146	147	175	180	973

Source: Grey & McDermott (1987).

The most significant development in terms of global FDI has been the rapid rise of Japanese FDI in the later 1980s, both acquisition and "greenfield" investments, in Europe and the US. By July 1989 the number of Japanese plants in Europe had more than doubled since 1983 to reach 411. (*Financial Times*, 28 June 1989, p.22). The UK had 92 such plants, France had 85, though the size of investment in the UK is much greater. French FDI has increased 70% to Ffr 34 billion in 1986, following the relaxation of controls on FDI by the Chirac government; about half went to the USA, and about 75% in total to the USA, Switzerland, Britain, Italy, The Netherlands and Federal Republic of Germany. Conversely, in 1985 100 French firms were acquired by foreigners, chiefly US (32) and UK (14) (Grey & McDermott, 1987).

Technology

Part of the increase in global activities of European, North American and Japanese firms is reflected in the transfer of technology from countries into the world system more generally. While it is not as straightforward a category to measure as financial flows, some indication of global trends is given by reference to patenting activity. Electronics is the most globalized of technologies in the terms discussed here, and the information in Table 4.8 refers only to the changing shares of key electronics patenting from 1969 to 1986.

Table 4.8 World patent shares in electronics, 1969–86 (%).

	USA		Europe		Japan		UK	
	69	86	69	86	69	86	69	86
Communications	72	58	15	20	7	20	7	4
Semiconductors	71	50	16	16	10	32	3	2
Image & sound	70	41	19	20	10	39	4	1
Computing	76	48	10	11	9	40	3	1

Source: Morgan et al. (1989)

Within Europe, the share of patenting activity by electronics firms also shows divergent trends over the period 1969–86, as Table 4.9 shows. It is clear that, on a global scale, a far greater share of patenting activity is now accounted for by Japan across the sectoral categories, though the USA still dominated in 1986. Europe has marginally improved its position, but has shown negligible growth by comparison with Japan.

Within Europe, the UK has lost the position of eminence it shared with

Technology

Table 4.9 European patent shares in electronics, 1969–86 (%).

	FR Germany		France		UK	
	1969	1986	1969	1986	1969	1986
Communications	21	27	16	20	32	15
Semiconductors	35	34	6	20	13	7
Image & sound	11	32	7	15	18	5
Computing	23	32	12	20	25	8

Source: Morgan et al. (1989)

FRG in 1969 and has moved down to the equivalent position to that of France at that date. While FRG has generally improved on its position between 1969 and 1986, France has made the greatest strides forward, pushing the UK from first or second most creative electronics economy to third across the board. Europe's strongest sector is communications, where the increase in patenting rate is greatest, and it is relatively strong in all three leading European producer countries. (Here, the UK has again fallen back, though not to a position as low as in other electronics fields.)

Regulation

Clearly, deregulation, and to some extent re-regulation, has played an important part in the globalization process, not just in telecommunications but in other sectors such as airline services, financial services and military industry. The removal of various national barriers to trade and foreign direct investment is unevenly developed, with the USA generally taking the lead, while Japan remains one of the most closed economies.

The mechanisms by which increased internationalization of interfirm and cross-market relations develop, in relation to regulatory policy changes, are activated by one or more of three policy shifts: deregulation, privatization and liberalization. If we take telecommunications as a paradigmatic case of a regulated utility, then the 1980s experiences of three countries – the USA, the UK and France – illustrate the manner in which hitherto sacrosanct national borders were breached.

Deregulation
In 1984 the US Bell system was dissolved, following decades of pressure upon AT&T (the regulated network manager of the US telecommunications service) by competitor companies on the one hand, and the courts

69

upholding anti-trust legislation on the other. The divestiture of AT&T assets, notably the regional Bell operating companies, left the former monopoly service supplier with four main business divisions: long-distance service; information systems (PABX exchanges, computing, etc.); technologies (componentry, networks, etc.); and R&D (Bell Laboratories) (Morgan & Pitt 1988). Because divestiture did not in fact amount to deregulation (since divestiture was imposed by the courts), AT&T remains regulated in long-distance service where it has faced competition from several new services. This has diminished AT&T's share of that particular market, causing the company to seek to develop its market potential overseas. This has been pursued in a major way by the company's information systems and technologies divisions, though for some time there have also been expectations that a European location would be sought for Bell Laboratories.

The approach of AT&T has been to form strategic alliances or joint ventures with firms in telecommunications switching, such as Philips through its APT joint venture (since joined by Italtel); computing, such as Olivetti (though that alliance has since been dissolved); components, as with the Spanish firm Telefonica; and software, with its recent purchase of the UK firm Istel. Thus, deregulation has drawn AT&T out into the global market at the same time as it has brought competition into the company's domestic market.

Privatization

In Britain, the telecommunications carrier was a nationalized corporation rather than a regulated network-management system. This was first separated from the Post Office and retitled British Telecom; then, in 1984 it was privatized and the service system was turned from a monopoly into a duopoly with the entry of Mercury Ltd into the telecommunications service market.

The opening up of the market had only a limited effect upon British Telecom's former monopoly, because of Mercury's small size and commitment to focus on business service, but it had a significant effect on the equipment suppliers. These had hitherto enjoyed a reasonably cosy, club-like relationship with British Telecom. But this arrangement was quickly changed by the opening up of the telecommunications equipment market to high-quality switch manufacturers such as Northern Telecom and Ericsson. The new telecommunications regime was now a regulated system, with Oftel having, by law, regulatory responsibility comparable to that of the Federal Communications Commission in the USA. While British

Telecom has sought to become active in overseas markets, its former suppliers have become more market-facing both domestically and overseas.

Each of the main suppliers has found great difficulty in managing the transition. First Plessey and GEC formed a joint venture in switching, then GEC and Siemens made a joint bid to acquire Plessey. This was successful and Plessey is gradually disappearing into the new mega-corporate structure. For technological reasons, STC was unable to get into the new generation of public switching. In its place, the company aimed for the decentralized switch market which, in theory, had useful compatibilities with its UK computing acquisition, ICL. However, neither STC nor ICL were large enough to compete in the new global era, and ICL has been acquired by Fujitsu while STC has been bought by Northern Telecom. Both companies previously had strategic alliances with STC-ICL.

Liberalization

In France, the telecommunications service remains publicly owned. However, the traditional public carrier was restructured in the 1970s and re-born as France Telecom. This company has been a highly successful public corporation, introducing both advanced services such as Minitel, and advanced switching and other technologies. As with Britain, the equipment-supply industry was a privileged group of companies that could historically rely upon public procurement providing a stable market. However, recent changes have seen some liberalization of the equipment-supply industry. Its restructuring first through nationalization, then by privatization, created a new structure of supply dominated by what has become a very large global corporation, Alcatel. However, other companies are in a competitive position in the liberalized equipment market, and new entries such as that based on the Matra–Ericsson alliance have made important inroads in specific market segments.

Thus, the rôle of regulatory or ownership change with respect to traditional modes of telecommunications service delivery have made a not insignificant, but by no means predominant, impact upon the globalization tendencies inherent in the industry. The main effects in the most radically transformed cases have been the diminution of domestic market share due to the incursions of new domestic or foreign entrants, and the pursuit of new markets overseas. Not only service-delivery companies but equipment suppliers have been significantly re-focused towards a globalized perspective by the combination of deregulation, re-regulation, privatization or liberalization in the past decade.

The globalization of C&C markets

It is clear that more and more economic activity – investment, trading, location, technology transfer and market penetration – is taking place on a global scale, nowhere more so than in computing and communications. It is also clear that the key corporations involved at the leading edge of the globalization process experience considerable unease as they seek to protect themselves from competitive pressures in their domestic markets, while stepping forward into new global markets with, sometimes, not a little trepidation. Uncertainty has been on the rise in all dimensions of corporate activity, and a major task of global players has been to try to control and reduce, at least at the margin, this burgeoning uncertainty. As, in many cases, governments have reduced regulatory responsibilities and actively pursued free-trade policies which often weaken hitherto "national champions", it is apparent that much of this uncertainty management must be undertaken by the firms themselves. Key efforts have been made to "keep tabs on" competitors, but also, more importantly, to share the costs, minimize the risks and lower the uncertainty of engaging in new research, development, production or marketing activities. Implementation has been by collaboration, partnership or strategic alliance formation with both complementary and, importantly, competitor companies (Cooke 1988).

In what follows, we will introduce some aspects of this process, since it transpires that a significant share of strategic activity undertaken by firms interviewed in the research reported here involved the formation of alliances. To this should be added more conventional acquisition, merger or joint-venture activity. We found, however, that much of this more conventional concentration behaviour was often preceded by alliance-building of a less conventional kind.

Alliances and acquisitions

Alliances – whether as a prelude to acquisitions, the development of joint-ventures or full-blown mergers – are driven by three market-related forces. First, for hardware producers, competition, whether from new entrants, from foreign producers of "clones" or from new products, has reduced the real selling price of their product, especially in computing, though observably so in telecommunications also. There is pressure to find new markets. This is difficult if they stay with the same products. So there is a desire to move into new product, or rather, service markets. This might involve a computer company in setting up a system for a client and supplying specialist software services in the process, as occurs commonly with financial trading networks. Alternatively, it might involve – as indeed it has for the two largest computing and communications firms in the world, IBM and AT&T – the attempt to get into the convergent technological field. A computer company enters telecommunications, or vice versa, so that the full network communication service can be supplied to the customer. Moving into new fields requiring new expertise, new markets, and products or services, over a relatively short time-period tends to make looking for a partner a very strong option, especially by comparison with going it alone.

A second alternative force encouraging alliance formation can be the need to get at a market not to sell product but to gain access to "know-how". A good example of this was the protectionism of the US government towards Japanese predatory purchases of US semiconductor manufacturers. Tariffs had been raised against Japanese imports and efforts had been made to get the US industry to help protect itself by setting up, *inter alia*, the Semiconductor Manufacturers Association. Faced with this, Fujitsu, wishing to purchase Fairchild, a Silicon Valley chip pioneer-firm in decline, found barriers to purchase being raised. Finally, Fairchild was sold to another Californian firm, National Semiconductor. Japanese firms thereafter sought to establish alliances and joint ventures with American microelectronics firms, so that they could maintain access to advanced know-how without unduly upsetting America's economic nationalists.

A third reason for alliance formation is the need to acquire a certain minimum scale of operations. Globalization means that the corporate groupings operating in the three main world markets of Europe, Japan and the USA have to be of a scale to compete, hypothetically at least, in each others' markets. Since the largest C&C firms have access to enormous resources for R&D, product development and marketing, these set the standard for the appropriate minimum scale of operations on a global scale. European firms have been the slowest to get up to the scale, though in the late 1980s there was a flurry of alliance, acquisition and merger activity to achieve this. The GEC–Siemens bid for Plessey was a paradigm case in point here, as was the earlier formation of Alcatel as a global telecommunications firm. However, large-scale and wide-ranging alliances may not be enough if the product range is too diffuse and occupying many of the same market segments as particularly powerful global competitors. The case of Philips is instructive in that, although it is one of Europe's largest electrical engineering companies, many of its alliance strategies (e.g. with AT&T and Siemens) were either very costly and in fields such as semiconductors where there is a huge gap to overcome in relation to Japanese producers, or where the partner had the superior technology, as for example AT&T switching, with which Philips could not compete. Moreover, many of Philips's traditional product lines are in markets where there is fierce competition from both outside and inside Europe. As a consequence, profitability has been seriously hit, large-scale redundancies have been announced and implemented, and the prospect of the company being broken up into separate units cannot be ruled out.

European Community policy

In the case of Europe, much of the collaborative activity by C&C companies has been initiated through EC programmes, notably ESPRIT, but such EC programmes generally have intra-European partnering as a requirement for funding. Nevertheless, ESPRIT has been reasonably successful in introducing C&C firms to each other around particular core technological research projects. Some of these have subsequently led to closer partnerships, joint-ventures, acquisitions and so on. The EC fear, justifiably, was that the tendency for European firms to look outwards to Japan and the US to acquire know-how and markets would lead to them being "hollowed-out" by their partners, losing ground in the technological race, and becoming bases through which non-European firms could market and distribute their product. The case of Philips in relation to AT&T is a good illustration of the real basis for this fear. Philips had developed an advanced digital telecommunications switch and saw the prospect of developing it to the leading position in the global market through gaining access to AT&T's world-class research facilities at Bell Laboratories. However, such was the scale, technological advantage, and sheer market power of AT&T that Philips engineers found themselves increasingly being employed as developers and adaptors of AT&T technology. This led to disaffection, the run-down of Philips's R&D capacity in telecom switching and the ultimate reduction of Philips's share in the APT joint venture to less than half its original 50%. Italtel now has a larger share in the AT&T International concern (as APT has been renamed) than Philips who were not prepared to act as AT&T's switching distribution outlet in Europe.

In research which has examined the success of ESPRIT in helping forge European research alliances (Mytelka 1990) it is shown how, over the 1980s, alliances within Europe have become more popular for three reasons: there is a chance, albeit slight, of gaining a technological advantage; specialized economies of scale can be retained because firms operate on more of an even playing field; and, collaborations are cheaper because of the ESPRIT funding, normally 50% of the cost of joint projects. In the early 1980s a core group of Bull, CGE, Philips, Thomson, Siemens and STET had 33 agreements with each other. This had grown to 44 by the late 1980s with a further 124 in existence between them and another six of the top twelve European firms: AEG, GEC, Nixdorf, Olivetti, Plessey and STC. Hence, a growing amount of intra-European alliance-building has taken place. Moreover, the relative weighting of partnering between European C&C firms has increased to rival the level attained by European and US firms. From 1983 to 1986 the latter rose from 32 to 49, while the former rose from 6 to 46.

Concluding remarks: globalization and corporate readjustment

It has been shown that globalization, though awaiting agreement on a precise conceptual definition, is a process having profound effects upon the strategic thinking of even the largest corporations in computing and communications. One of the key changes that justifies the use of the term globalization (in preference to the more traditional term inter-nationalization) to describe the process under discussion is that, in the first two-thirds of the present century, corporations with global reach tended to originate from a few national economies, notably the US and the UK. Now there are far more such corporations originating in more national economies (Japan, to some extent Korea, and more European countries, as well as more US companies). Moreover, the largest companies such as IBM, Sony, Alcatel and so on see themselves as global corporations without necessarily appending a particular national significance to their location of origin. This is clearest in the case of the largest Japanese corporations, many of which profess a strategy of "global localization". This means establishing within each of the three main world market areas of the USA, Europe and South-East Asia, structures of production that are complete from the conception (R&D) to the marketing stage and which are "locally assimilated". The process by which this is intended to be achieved is summarized in Masuyama (1990) in a procedure described as the "road to globalization" for Japanese corporations. This involves the following steps:

1 establishment of marketing offices
2 raising the ratio of local production
3 establishment of overseas financing offices
4 establishment of R&D facilities
5 promotion of locals to managerial positions
6 independence of management by local headquarters
7 public offering of stock of locally incorporated companies
8 diversification of business

At each step in the process, the objective of enhancing local assimilation is given high priority. This is an important difference from the traditional process of internationalization which made no strong claim to have the objective of local assimilation, something which may well prove to have been one of its prime weaknesses.

One of the crucial justifications for global localization is the overwhelming requirement in highly competitive markets to meet customer need. As it has been recently put by a leading commentator on corporate readjustment in the context of globalization and partnership formation:

"The name of the game in most industries was simply beating the competition. If you discovered an ounce of advantage, you strengthened it with a pound of proprietary skill or knowledge. Then you used it to support the defensive wall you were building against competitors.

The forces of globalization turn this logic on its head. You can't meet the value-based needs of customers . . . entirely on your own. You can't do without the skills and technology of others. You can't even keep your own technology to yourself for long. Having a superior technology is important, of course, but it is not sufficient to guarantee success in the market. Meeting customer need is the key . . ." (Ohmae 1989)

Hence, strategic alliances become competitive weapons in a globalizing economy. Interest in this particular aspect of corporate readjustment has yielded a significant increase in research literature on the subject (Harrigan 1985a, Wiseman 1985, Hacklisch 1986, Perlmutter & Heenan 1986, Porter & Fuller 1986, Teece 1986, Ernst 1987, Von Hippel 1987, Chesnais 1988, Hagedoorn & Schakenraad 1988, Pisano et al. 1988, Hagedoorn 1989, Mytelka 1990). As Chesnais (1988) puts it:

". . . strategic alliances are formal and informal agreements between two or more companies providing for a certain degree of collaboration between them and involving equity and non-equity participation or the creation of new companies."

Vague though this definition remains, it is worth keeping in mind the extent to which such activities have increased in number in C&C industries. For example, according to data held in the MERIT databank (for example, see Hagedoorn & Schakenraad 1988) the firms with the top ten strategic partnerships in information technologies over the period 1985 to 1989 were as follows: Siemens (134), Philips (127), Olivetti (110), IBM (108), Hewlett–Packard (96), DEC (95), AT&T (90), Thomson (83), Fujitsu (78) and Motorola (68). Clearly, for these firms, the phenomenon has a reasonably important rôle to play in their strategic thinking as global corporations.

To summarize the rationale for this by way of a simple conceptual frame (Teece 1986, Hagedoorn & Schakenraad 1988), it may be proposed that the options for firms confronted by the kind of fierce global competition, deregulation and market uncertainty under discussion, are threefold: seek control, pursue stand-alone competition, or collaborate. Control has been the Fordist strategy, whereby elements of necessary business activity are absorbed into the firm through merger or acquisition. This method of acquiring what Pisano et al. (1988) refer to as "complementary assets" enhances "strategic competence". This is particularly so where the complementary asset being sought is specialist, as is normally the case with advanced technology. This option remains a major strategic practice, though, as we have seen with IBM's ventures into telecommunications technologies and AT&T's into computing, such decisions can come seriously unstuck.

Normally, the alternative to exerting control through ownership has been to acquire the necessary complementary asset by market transaction carried out in the normal, competitive-bidding posture. Where the required asset is widely available on the market and is, further, a difficult service or product to absorb into the corporate production structure, it is highly likely to be made the subject of a normal customer–supplier contractual relationship.

However, where neither of these strategic options is open to the firm in question, alliance formation may follow. Collaboration may arise because, even if the asset sought is non-specialist, it may be difficult to absorb owing to patenting or other intellectual property law. Or, it may, of course, be an asset such as access to a foreign market protected by tariff or non-tariff barriers. In these circumstances the alliance, agreement or partnership of a contractual or non-contractual kind will be the likely strategic option pursued. Collaboration clearly does not rule out competition, as we have seen; indeed it is often seen as a partial alternative form of continuing the competitive struggle. Each member of the partnership seeks, albeit temporarily perhaps, to benefit from the other's competences to improve market strategy.

Finally, it should be noted that this particular strategy need not necessarily function with equal value for both partners. Most likely, one will benefit from the partnership more than the other. This can be the prelude to withdrawal from the alliance, as has been noted earlier. Collaboration adds to management costs, introduces an asset or assets to a possible new competitor, and obviously diminishes control for both partners. The advantages have already been noted: access to new technology, new markets, reduction of marketing costs, for example. Some commentators (e.g. Hamel et al. 1989) have drawn attention to the downside of the alliance strategy as an explanation for the brevity of the experience for many companies. Nevertheless, more companies are engaging in a larger number of such links, and in the process are moving up the alliance-management learning curve. Of the many firms interviewed in this research who found alliance building to be advantageous, one such, Ericsson, saw "value-added partnering" as extremely valuable, provided care was taken over choice of partner, as part of a broader strategy of seeking to manage the globalization process.

CHAPTER FIVE

The regional patterns of computing and communications industries in the UK and France

Erik Swyngedouw
Martine Lemattre
Peter Wells

This chapter presents a comparative analysis of the regional structure of computing & communications equipment and component production, as well as service provision. In the first section, some theoretical considerations on the location pattern of high-technology industries will be offered. Next, the methodology of the research is described, and definitions of telecommunications and computer-related subsectors are explored. Then, the regional pattern of these subsectors in the UK and France is examined. Later, the qualification structure of the labour force is further explored. We conclude with some tentative suggestions concerning the dynamics of regional growth based on high technology.

Identifying the information industry in France and the UK

If we take a very broad definition of information activities and processes, it would include the following: a process of emission, transformation and transmission of information. This definition is reflected by three broadly defined economic sectors: informatics (computers), electronics and tele-communications, in association with their respective support services. In fact, for each of these activities, both the technological divides as well as the manufacturing/services cleavage are becoming increasingly irrelevant. The drive towards the integration of systems, which puts a premium on technological convergence and product symbiosis, blurs the traditional distinction between sectors of industrial activity, while the service–manufacturing relationship is becoming more complex and mutually interdependent. On the one hand, software engineering, systems-integration services and other similar activities are integral parts of the manufacturing of information systems, while on the other hand the production of "hardware" is influenced and sometimes even conditioned by the development of "software", both in its real meaning (strings of computer language instructions) as well in a more generic sense (systems-integration consulting and engineering services). These considerations suggest that a classical sectoral analysis of spatial-economic dynamics may not be the appropriate means of disentangling the intricate relationship between technologies which are apparently widely divergent, and production processes and products. Elsewhere in this book, these observations are analysed in greater detail in both theoretical and practical terms.

Unfortunately, national or regional aggregate employment or other data are available only on a sectoral basis. To complicate matters even further, neither the UK Standard Industrial Classification (SIC) nor the French "Nomenclature des Activités et des Produits" (NAP) are attuned to take into account changes in the nature of either production processes or product ranges. Moreover, international comparative analysis is complicated by the fact that the principles of industrial classification or of data collection are not necessarily consistent across countries. Indeed, for France and the UK, sectoral definitions are not always the same and they may show more or less important differences and divergences. These issues make a comparative and dynamic analysis of sectoral and spatial changes in information activities rather difficult. Nevertheless, an attempt will be made to identify the grand sectoral–spatial changes in the two countries.

With these provisos in mind, we identified for each country the sectors directly involved in the production and servicing of information and telecommunications systems. For the UK, 10 four-digit SIC sectors were identified. For the French case, 16 NAP-600 (sub)sectors were identified. For both countries, the sectors include the servicing industries, the systems-integration industries and the equipment manufacturers, as well as the producers of active electronic components. As can be seen from Table 5.1 (which illustrates similarities as well as major discrepancies), the differences between the two countries are quite substantial. Even in cases with apparent similarities, the classification procedures used by the respective statistical offices in the UK and France may result in quite important differences.

Table 5.1 The identification of information technology subsectors in the UK (4-digit SIC classification) and in France (NAP-600 classification).

SIC Name		NAP-600 Name	
3301	Electrical instruments	2815	Equipements d'automatisation de processus industriels
		2912	Appareils de radiologie et d'électronique médicale
		2913	Appareils de contrôle et de régulation
3302	Electronic data-processing equipment	2701	Materiel de traitement de l'information
3441	Telecoms, equipment	2911	Materiel télégraphique et téléphonique
3442	Office machinery	2702	Machines de bureau
3443	Radio/electronic capital goods	2914	Materiel professionnel électronique et radio électrique
3444	Components: electronic equipment	2915	Composants passifs et condensateurs fixes
3453	Active components /sub-assembly	2916	Tubes électroniques et semi-conducteurs
3454	Electronic consumer goods /(other)	2921	Appareils radio-récepteurs et téléviseurs
		2922	Appareils d'enregistrement et de réproduction(son/image)
7902	Telecommunications services	75	Services de télécommunication et postes
8394	Computer services	7704	Travaux à facon informatiques
		7703	Cabinets d'études informatiques et d'organisation
		7702	Cabinets d'études économiques et sociologiques
		7701	Cabinets d'études techniques

Recent trends in C&C industries in the UK and France

Tables 5.2a&b show the national trends for each of the selected subsectors in the two countries. In UK information-manufacturing sectors, employment in electronic data-processing equipment, office machinery and electronic consumer goods rose significantly, while others, and most notably the telecommunications equipment producers, suffered major job losses over the 1981–87 period. Moreover, the employment base of both electronic component production and electrical & electronic equipment production were equally eroded. Employment in telecommunications services virtually stagnated during the 1980s, while computer services employment rose spectacularly to see its employment base almost double between 1981 and 1987.

The French data reveal similar patterns. In the manufacturing sectors, the telecommunications industry lost many jobs, together with the more mature electronics subsectors, but data-processing equipment and semiconductors saw their employment level rise. Employment in information services increased in a quite dramatic way. Both countries show important changes in organization of the key industries. Moreover, growth is further stimulated by the convergence of technology and services. "Global systems solutions", as well as the necessary support services, are becoming key value-adding elements in the information-systems production process.

Table 5.2 Total employment evolution

(a) UK information technology and service industries

SIC	Name	1981	1984	1987	% change 1981–87
3301	Electrical instruments	18,768	15,933	13,344	-29
3302	Electronic DP equipment	56,799	71,225	70,230	24
3441	Telecoms equipment	58,632	47,248	30,942	-47
3442	Office machinery	26,746	34,418	38,035	42
3443	Radio/electronic capital goods	88,761	73,292	70,916	-20
3444	Components: electronic equipment	29,735	29,290	25,705	-14
3453	Active components /sub-assembly	75,654	64,712	57,897	-23
3454	Electronic consumer goods, etc.	44,816	56,361	64,611	44
7902	Telecoms services	234,089	229,723	224,350	-4
8394	Computer services	54,722	78,699	108,416	98

Source: Department of Employment.

(b) French information technology industries

NAP-600	Name	1975	1983	% change 1975–83
2701	Materiel de traitement de l'information	41,207	51,520	25.0
2702	Machines de bureau	3,997	3,793	-5.1
2815	Equipements d'automatisation de processus industriels	3,155	10,494	232.6
2911	Materiel télégraphique et téléphonique	74,780	63,648	-14.9
2912	Appareils de radiologie et d'électronique medicale	4,933	4,645	-5.8
2913	Appareils de contrôle et de régulation	31,247	24,556	-21.4
2914	Materiel professionel électronique et radio électrique	61,885	65,668	6.1
2915	Composants passifs et condensateurs fixes	27,263	23,456	-14.0
2916	Tubes électroniques et semi-conducteurs	10,853	18,270	68.3
2921	Appareils radio recepteurs et téléviseurs	18,193	16,056	-11.75
2922	Appareils d'enregistrement et de réproduction (son/image)	7,600	6,093	-19.8
75	Services télécommunication et postes	102,318	127,033	24.0
7701	Cabinets d'études techniques			
7702	Cabinets d'études économiques et sociologiques			
7703	Cabinets d'études informatiques at d'organisation	19,036	40,639	113.0
7704	Travaux à façon informatiques	20,241	31,471	55.0

Source: INSEE, Lemattre M.

Regional structure and evolution of employment

In what follows, the regional structure and dynamics of some of the key information-technology and service industries will be explored. Given the differences in statistical definitions and concepts between France and the UK, a slightly different classification is used for the two countries. The geographical basis on which the data are collected also differs significantly. For France, the region is the basic geographical unit for our analysis. For the UK, travel-to-work areas have been used. However, despite those statistical differences, some remarkable similarities highlight the global and more general dynamics of recent changes in the IT and service sectors, while at the same time national differences will be identified.

Table 5.3 lists the sectors used in the UK and France respectively. For the UK account, four sectors will be examined more closely: electronic data-processing equipment, telecommunications equipment, telecommunications services and computer services. For France, five aggregated sectors will be further analysed: office material (including computers), electronic components, telecommunications equipment, electronic equipment production and information services (including telecommunications and computer services).

The regional structure of information-equipment production and services in the UK

Between 1981 and 1984, the electronic data-processing industry grew by more than 26% in employment, but this increase turned into a decline of 1.4% between 1984 and 1987. The sector is massively concentrated in the South East, with 47.1% of total employment in that region in 1987. Scotland shows the second largest concentration with 13%, illustrating the effect of both the Silicon Glen phenomenon and of the concentration of the offshore oil industry in the Aberdeen area. London has 12% of total employment in this sector. The North West and the West Midlands are two other regions with significant employment concentrations, respectively 9.5% and 7.1% of total employment (see Table 5.4).

In terms of regional trends (see Table 5.4 and Figs 5.1 & 2), a number of important observations can be made. First, the South East saw employment continuing to grow, in both relative and absolute terms, strengthening its overall position. In absolute terms, London, on the contrary, grew only in the first period, and shrank in the subsequent period. Overall, London lost some of its employment share in favour mainly of the rest of the South East. While in the period 1981–84, both regions experienced a positive growth, the period 1984–87 was characterized by a relative displacement of employment and by a marked tendency towards decentralization away from London. This decentralization affected not only the South East but also the outer regions such as the West

Table 5.3 Identification of the core information-technology and service industries in the UK and France.

United Kingdom		France	
SIC	Name	NAP-600	Name
3302	Electronic DP equipment	2701	Materiel de traitement de l'information
		2702	Machines de bureau
		2915	Composants passifs et condensateurs fixes
		2916	Tubes électroniques et semi-conducteurs
3441	Telecoms equipment	2911	Materiel télégraphique et téléphonique
		2912	Appareils de radiologie et d'électronique medicale
		2913	Appareils de contrôle et de régulation
		2914	Materiel professionel électronique et radio électrique
7902	Telecoms services	75	Services télécommunication et postes
		7701	Cabinets d'études techniques
		7702	Cabinets d'études économiques et sociologiques
		7703	Cabinets d'études informatiques et d'organisation
		7704	Travaux à façon informatiques
8394	Computer services		

Table 5.4 Employment change in EDP equipment, 1981–87 (SIC 3302).

Region	1981	%	1984	%	1987	%
South East	25,402	44.7	32,750	46.0	33,109	47.1
East Anglia	466	0.8	1,124	1.5	766	1.0
London	8,840	15.5	9,168	12.9	8,430	12.0
South West	2,160	3.8	1,583	2.2	·1,496	2.1
W. Midlands	3,698	6.5	4,406	6.1	5,430	7.7
E. Midlands	831	1.4	2,235	3.1	1,131	1.6
Yorks & H'side	792	1.4	1,544	2.1	1,235	1.7
North West	6,012	10.5	5,619	7.9	6,690	9.5
North	338	0.6	139	0.2	234	0.3
Wales	1,179	2.0	2,748	3.8	2,563	3.6
Scotland	7,086	12.5	9,875	13.8	9,133	13.0
TOTAL	56,804	99.7	71,191	99.6	70,217	99.6
Change 1981–87	*+114,387(+26.0%)*			*−998 (−1.4%)*		
	+13,389 (+24.3%)					

Table 5.5 Employment change in telecoms equipment, 1981–87 (SIC 3441)

Region	1981	%	1984	%	1987	%
South East	5,269	9.0	2,718	5.7	2,990	9.6
East Anglia	350	0.6	312	0.6	367	1.1
London	12,068	20.5	8,602	18.2	2,233	7.2
South West	665	1.1	1,817	3.8	1,371	4.4
W. Midlands	14,762	25.1	13,249	28.0	9,373	30.2
E. Midlands	5,894	10.0	4,714	10.0	3,830	12.3
Yorks & H'side	358	0.6	746	1.6	1,154	3.7
North West	7,110	12.1	5,971	12.6	4,320	14.0
North	5,621	9.5	4,048	9.0	1,582	5.1
Wales	3,711	6.3	3,092	6.5	2,663	8.6
Scotland	2,825	4.8	1,980	4.1	1,059	3.4
TOTAL	58,633	99.6	47,249	100.0	30,942	99.6
Change 1981–87	*−11,384 (−19.4%)*			*-16,307 (-34.0%)*		
	−27,691 (−47.0%)					

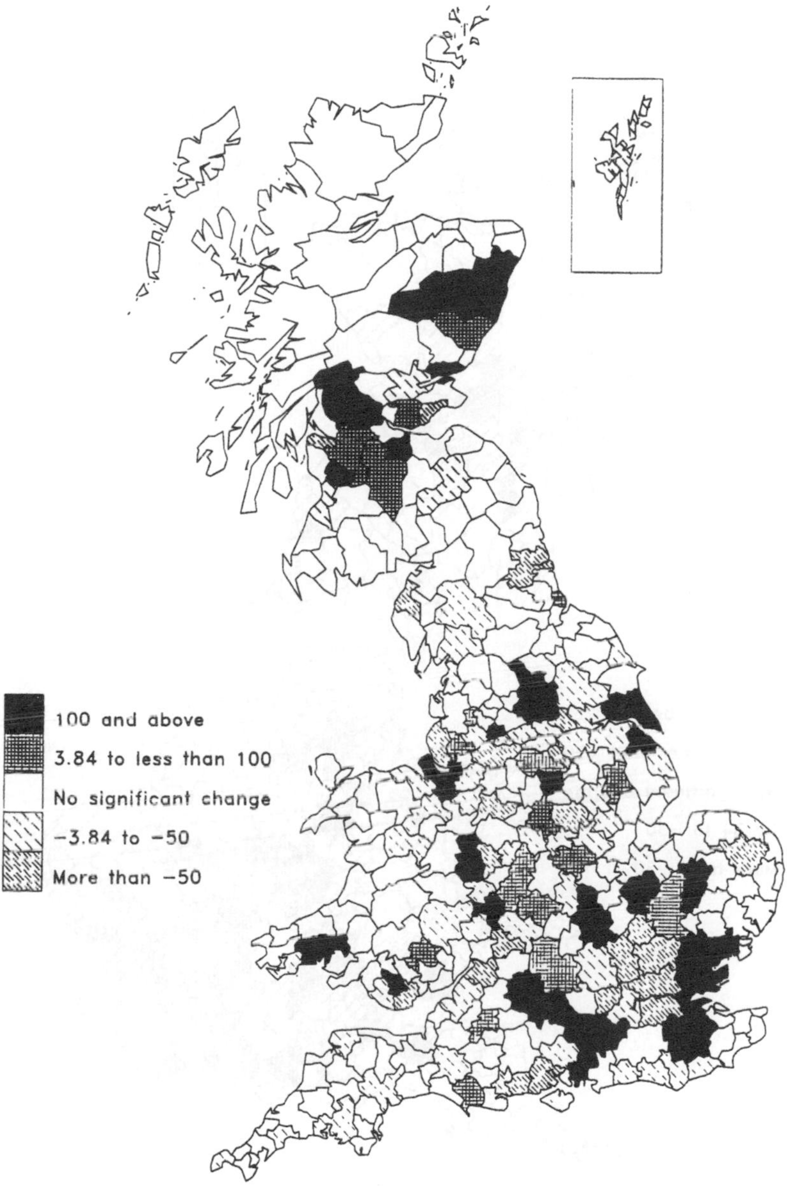
100 and above
3.84 to less than 100
No significant change
−3.84 to −50
More than −50

Figure 5.1 Employment in electronic data processing, 1981–84
Change measured in chi-squared; travel-to-work areas, 1984.
(*Source:* Department of the Environment statistics [NOMIS])

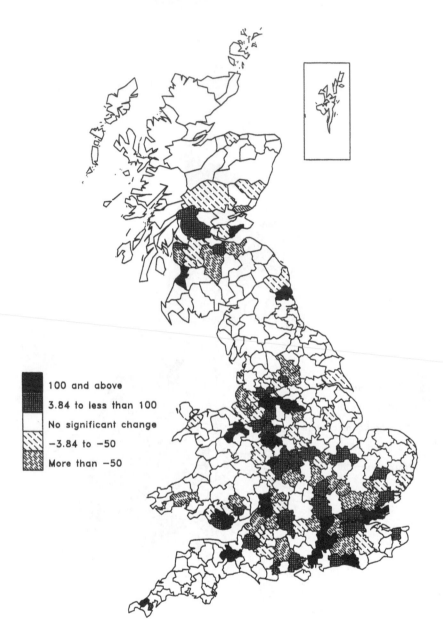

Figure 5.2 Employment in electronic data processing, 1984–87
Change measured in chi-squared; travel-to-work areas, 1984.
(*Source:* Department of the Environment statistics [NOMIS])

Legend:
- 100 and above
- 3.84 to less than 100
- No significant change
- −3.84 to −50
- More than −50

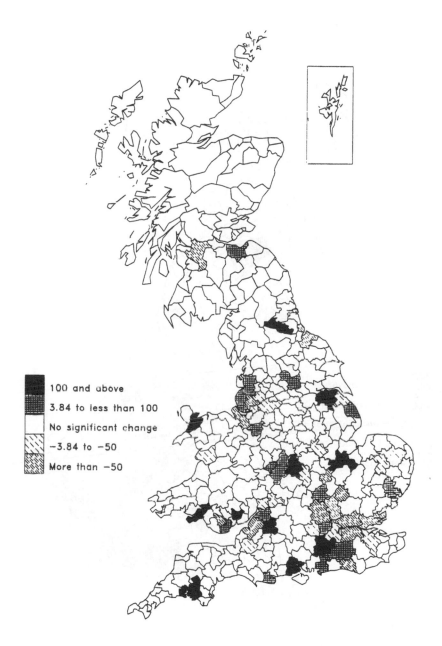

Figure 5.3 Employment in telecommunications equipment, 1981–84
Change measured in chi-squared; travel-to-work areas, 1984.
(*Source:* Department of the Environment statistics [NOMIS])

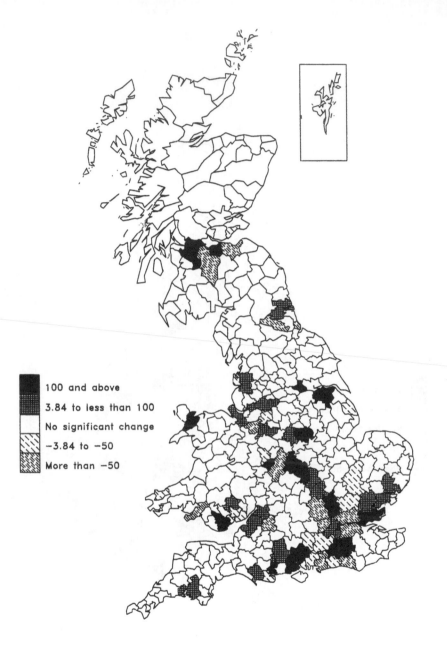

Figure 5.4 Employment in telecommunications equipment, 1984–87
Change measured in chi-squared; travel-to-work areas, 1984.
(*Source:* Department of the Environment statistics [NOMIS])

Legend:
- 100 and above
- 3.84 to less than 100
- No significant change
- −3.84 to −50
- More than −50

Midlands and the North West; and, to a lesser extent, South Wales benefited from this decentralization during a period of overall decline in the total number of jobs in the sector. As will be demonstrated below, this process, which accelerated after 1984, followed the pattern of decentralization in telecommunications equipment production, but with a certain time-lag. Besides the obvious reasons for decentralization, processes of industrial restructuring and reorganization contributed to this movement. It is possible to understand much of this regional shift in terms of the arrival from foreign sources of direct investment in non-metropolitan locations, and the relative decline of indigenous electronic data-processing firms.

There may be some connection between the outward shift in electronic data-processing and the preceding outward move of telecommunications equipment (see Table 5.5). The strongest feature here would be the weakness of UK industry, with its attachment to less advanced technologies, the production of which had traditionally been in large cities such as London. The fact that computing is also a communications technology, and is thus a competitor to telecommunications equipment, may also have contributed to the fall in total employment in the telecommunications industry while the computer industry continued to expand. Certainly, some computer manufacturers have increasingly moved towards the integration of the production of communications-related equipment in their product range. The decline of employment in the telecommunications sector, therefore, has to be evaluated in the light of the rise employment in computer production. The telecommunications equipment industry lost almost half its total employment over the period between 1981 and 1987, suggesting (a) a major contraction of telecommunications activities in a context of growing telecommunications equipment output and (b) a blurring of traditional sectoral divides, which makes it increasingly difficult to distinguish functionally between the various information emission, transformation and transmission technologies.

The decline in telecommunications equipment employment is highly uneven in geographical terms. London experienced a dramatic reduction of its employment bases in this sector, losing almost 10,000 jobs. From a position of relative dominance in 1981 (20.5%), London's share fell to 7.2.% in 1987. This process was particularly pronounced during the period 1984–87. This dramatic decrease was not matched by a subsequent rise of employment in the South East. In fact, in absolute terms, the South East lost about half of its employment and only just maintained its share in the national total (9.6%). Most other areas experienced a decline in employment, with the exception of those regions whose share had been very low at the beginning of the eighties, such as Yorkshire–Humberside and the South West, which improved their relative positions quite remarkably. This process seems to suggest that decentralization away from the core regions continued through the 1980s. The West Midlands, for example, which already had a high percentage of telecommunications employment in 1981 (25.1%) actually saw their share go up to 30.2% despite a loss of more than 5,000 jobs. The shares of the East Midlands (12.3%) and the North West (14%) are equally high. This relative geographical concentration is associated with the dominance of GEC–Plessey (GPT) as the main British public-switching telecommunications equipment producer. GPT's production facilities are mainly concentrated in these three regions. Wales saw its improving position in high-technology production confirmed and it strengthened its share in total employment, reaching 8.6% in 1987, despite a loss of 1,000 jobs over this period. Figures 5.3 & 4 detail the regional dynamics of telecommunications equipment production and show both the relative spatial concentration of the sector as well as the uneven effects associated with the decline of employment.

Table 5.6 charts the regional evolution of employment in computer services. Employment in this sector has virtually doubled over the 1981–87 period, totalling 108,418 workers in 1987. Moreover, the growth process took place in all regions, without massive changes in the relative patterns of regional concentration. Nevertheless, the process of decentralization, although perhaps less pronounced than in the case of the manufacturing sectors, is also observable in this sector. However, the spatial scope of the process is rather limited. Indeed, London lost its dominant position, falling from a 32.2% share in 1981 to 28.4% in 1987. The rest of the South East, on the contrary, reversed its position and became the leading region in terms of employment concentration in computer services. Its relative share jumped from an already high 24.1% in 1981 to 29.4% in 1987. Other regions basically remained at the same level, with the possible exception of the West Midlands, which lost 2% of its share, and East Anglia, which increased its share from 1.7% to 3.1%. The chi-square maps (Figs 5.5 & 6) document this decentralization, away from both London and the relative growth of the belt around London, extending to South Wales along the M4 corridor. As in the case of the computer industry, marked increases can also be observed in Scotland (Silicon Glen and the Aberdeen area). This concentrated growth of computer services in Scotland is closely

Table 5.6 Total employment in computer services, 1981–87 (SIC 8394).

Region	1981	%	1984	%	1987	%
South East	13,219	24.1	21,216	26.9	31,918	29.4
East Anglia	935	1.7	1,716	2.1	3,460	3.1
London	17,659	32.2	23,665	30.0	30,865	28.4
South West	3,133	5.7	4,707	6.0	6,166	5.6
W. Midlands	5,990	10.9	7,782	9.8	9,635	8.8
E. Midlands	1,261	2.3	2,081	2.6	3,985	3.6
Yorks & H'side	3,217	5.8	3,767	4.7	4,792	4.4
North West	6,150	11.2	8,299	10.5	11,243	10.3
North	662	1.2	1,093	1.3	1,221	1.1
Wales	813	1.5	1,594	2.0	1,548	1.4
Scotland	1,684	3.0	2,779	3.5	3,585	3.2
TOTAL	54,723	99.6	78,699	99.4	108,418	99.3
Change	*+23,976 (+43.8%)*			*+29,719 (+37.7%)*		
1981–87			*+53,695 (+99.1%)*			

associated with the development of the high-technology manufacturing industry, whose growth is in part linked to hardware and services–software development. The favourable development of the Aberdeen area stems from the specialist demand for computer services coming from offshore oil production.

Both the recent evolution and regional distribution of telecommunications services remained fundamentally unchanged over the 1981–87 period (Table 5.7). Indeed, they are fundamentally demand-led and therefore they more or less follow the spatial distribution of households and businesses. It is little surprise, therefore, to find that the spatial distribution of telecommunications services resembles the spatial structure of population and of telecommunications-intensive economic activities. This is confirmed by the relatively important shares of Scotland (7.3%), where Silicon Glen, in particular, generates an important demand for high-level telecommunications functions (see also Figs 5.7 & 8).

In recent years, however, a tendency of decentralization from London can be observed. The share of London in total telecommunications-services employment fell from 33.8% in 1981 to 31.2% in 1987. East Anglia saw its share rise from only 1.3% in 1981 to 4.7% in 1987, while the South East increased its share from 16.3% to 17.8%. British Telecom, for example, has aggressively decentralized its telecommunications services away from London, mainly for cost reasons. Mercury (BT's competitor in the telecommunications-services market) is overwhelmingly business- rather than household orientated. The exploitation of this particular high-value-added market niche leads to concentration, given the demand-led nature of telecommunications services. Mercury originally centralized in London, but subsequently decentralized away from London towards the suburban belt. Such forms of focused service decentralization and associated radial spatial movements are much more difficult for providers of consumer-orientated telecommunications service providers.

However, although there still is a strong relationship between services, population and infrastructure, the nature of the new technological systems profoundly affects the spatial structure of telecommunications service centres. For example, large-scale digital networks reduce the need for small exchanges. This leads to spatial centralization. However, this centralization is accompanied by concentrated decentralization out of London for *cost* reasons, exemplified by the location of Racal (servicing mobile telecommunications) in Basingstoke.

Table 5.7 Total employment in telecoms services, 1981–87 (SIC 7902.)

Region	1981	%	1984	%	1987	%
South East	38,310	16.3	38,931	16.9	39,946	17.8
East Anglia	3,213	1.3	9,224	4.0	10,660	4.7
London	79,299	33.8	74,897	32.6	70,183	31.2
South West	16,770	7.1	16,695	7.2	15,856	7.0
W. Midlands	19,421	8.2	17,949	7.8	17,252	7.7
E. Midlands	8,855	3.7	10,502	4.5	9,189	4.0
Yorks & H'side	13,077	5.5	13,638	5.9	13,103	5.8
North West	17,097	7.3	15,972	6.9	17,235	7.6
North	7,663	3.2	7,819	3.4	6,969	3.1
Wales	7,595	3.2	8,562	3.7	7,588	3.3
Scotland	18,792	8.0	15,530	6.7	16,571	7.3
TOTAL	230,092	97.6	229,719	99.6	224,552	99.5
Change			*-373 (-1.6%)*		*-5,167 (-2.2%)*	

Changes in the composition of skills in the key C&C sectors are rather more difficult to identify. Nevertheless, data supplied by the Engineering Industry Training Board (EITB) (which excludes telecommunications services and computer services) does underline the qualitative advantage enjoyed by the South East. Table 5.8 shows the proportion of the total workforce in categories 1 & 2 (managers, professional, engineering scientific and technical staff) aggregated for eight SICs (i.e. those in Table 5.2a excluding SIC 7902 Telecommunications Services and 8394 Computer Services).

Clearly, there has been a generalized aggregate shift in the composition of the workforce, the two elite categories having increased their share of total employment from 15.9% to 20.7% between 1983 and 1989. However,

94

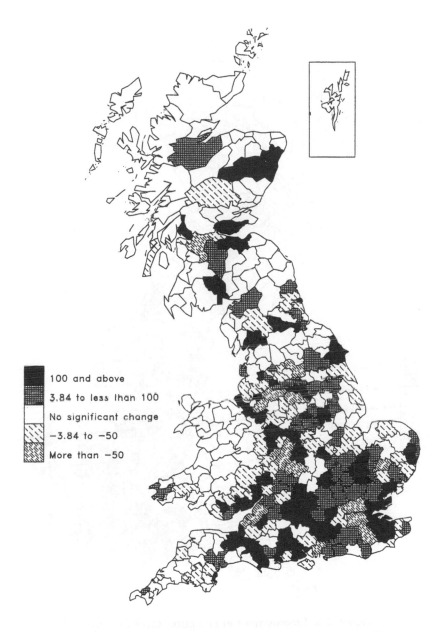

Figure 5.5 Employment in computer services, 1981–84
Change measured in chi-squared; travel-to-work areas, 1984.
(*Source:* Department of the Environment statistics [NOMIS])

Legend:
- 100 and above
- 3.84 to less than 100
- No significant change
- −3.84 to −50
- More than −50

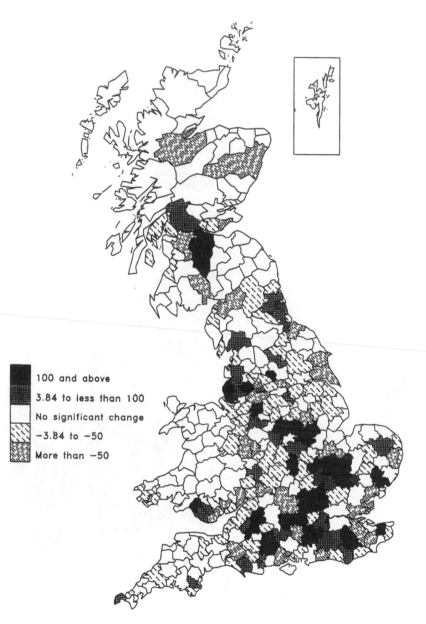

Figure 5.6 Employment in computer services, 1984–87
Change measured in chi-squared; travel-to-work areas, 1984.
(*Source:* Department of the Environment statistics [NOMIS])

Legend:
- 100 and above
- 3.84 to less than 100
- No significant change
- −3.84 to −50
- More than −50

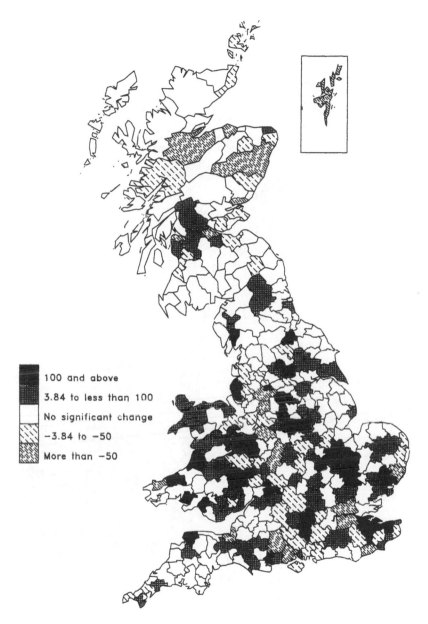

Figure 5.7 Employment in telecommunications services, 1981–84
Change measured in chi-squared; travel-to-work areas, 1984.
(*Source:* Department of the Environment statistics [NOMIS])

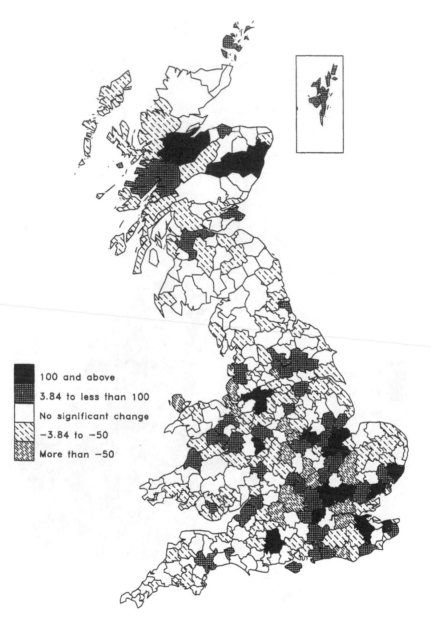

Figure 5.8 Employment in telecommunications services, 1984–87
Change measured in chi-squared; travel-to-work areas, 1984.
(*Source:* Department of the Environment statistics [NOMIS])

Legend:
- 100 and above
- 3.84 to less than 100
- No significant change
- −3.84 to −50
- More than −50

some erosion of the dominance of the South East is also evident in terms of these elite employment categories. While in 1983 the South East was the only region to record elite employment levels above the country's average, in 1989 it was joined by the West Midlands and the North West. Wales and the northern regions remained significantly below the country's average.

Table 5.8 Management & professional, engineering, scientific and technical staff in information-equipment industries (% of total workforce).

Region	1983	1989
South East	18.2	23.9
East Anglia	15.2	16.1
South West	15.3	20.5
West Midlands	15.1	23.2
East Midlands	17.1	21.9
Yorkshire & Humberside	14.0	15.3
North West	13.2	22.3
North	5.1	7.3
Wales	7.8	9.7
Scotland	15.3	18.2
TOTAL	15.9	20.7

The regional structure of information-equipment production and services in France

The spatial distribution of the information industry in France

This section discusses first the spatial distribution and dynamics of the information-equipment industry: the total of manufacturing NAP-600 sectors as defined in Table 5.2b (excluding information services). These data are summarized in Table 5.9 and Figures 5.9 & 10.

The dominant geographical feature of the information industry in France is the weight of the Paris region. This supremacy, however, has been slightly diminishing over the past decade or so. In fact, the region Île de France, although still accounting for 41.8% of total employment in those sectors in 1987, is actually losing jobs in contrast with most other regions. In overall terms, information industries gained employment between 1975 and 1981, but a period of decline set in after 1981, a process which became more pronounced during the 1983–87 period. The Paris region, although maintaining its dominant position, lost more than 24,000 jobs, and its relative share fell from 49.8% in 1975 to 41.8% in 1987. Two major areas of relatively important concentration can also be distinguished further. The regions around Île-de-France (Région Parisienne) show a relatively significant concentration of employment in information industries (Centre, Pays de la Loire, Haute Normandie, Basse Normandie and Bretagne). Their evolution, however, is relatively uneven. The regions west of Paris (Centre, Pays de la Loire and Bretagne) increased their relative share over the 1975–87 period, while the others experienced decline in varying degrees. The second important area is the South of France. Relatively important concentrations of employment can be found in Rhône–Alpes (7.2%) and Provence–Côte d'Azur (PACA) (4.1%), while all southern regions grew rapidly over the 1975–87 period. In short, a double process of concentrated decentralization away from Paris can be observed in France. On the one hand, there is a marked tendency for decentralization towards the immediate periphery of Paris, extending into Brittany in the west, while, on the other, major growth and concentration of information activities is taking place in southern France.

The decentralization of the C&C industry

The above description of the geographical distribution and tendencies of the information industries hides a series of more complex realities. In particular, the relationship between economic decentralization and geographical displacement on the one hand, and the important rôle of national decentralization policies in the electronics sector on the other, must be more closely examined in terms of their effects on particular regions.

Three distinct phases can be identified in the development of the information industries in time and space. The first is the origin of the electronics industry in the Paris region. Then, from the mid-1950s, these industries underwent a process of accelerated decentralization, resulting in a more or less important development of information industries in peripheral regions. Finally, since the mid-1970s, the restructuring process has reshuffled the map in quite radical ways. The process of decentralization towards designated "target" regions came to a rather abrupt end, while development in the southern regions accelerated.

In the present discussion, we focus on this most recent period, but it is useful to recall briefly the process of decentralization in France since the mid-1950s (see Durand 1974). Originally, between 60% and 90% (depending on the particular sector) of electronics-related employment was concentrated in the Paris region. The only noticeable exception to this general rule was the prominence of the area around Grenoble in the Rhône–Alpes region. A combination, then, of favourable economic conditions and the systematic attention paid to the expansion of the

Table 5.9 Change in the regional distribution of employment in the French information industry.

Regions	TOTAL					TOTAL (%)				
	1975	1978	1981	1983	1987	1975	1978	1981	1983	1987
Région Parisienne	141,900	140,742	140,118	129,554	117,312	49.80	48.70	47.95	44.96	41.83
Rhône–Alpes	14,450	16,352	18,245	18,722	20,360	5.07	5.66	6.24	6.50	7.26
Pays de la Loire	19,295	20,416	20,390	21,089	19,548	6.77	7.06	6.98	7.32	6.97
Centre	14,868	18,449	17,005	18,098	17,933	5.22	6.38	5.82	6.28	6.39
Bretagne	13,152	8,658	13,527	14,224	14,310	4.62	3.00	4.63	4.94	5.10
PACA	8,260	10,210	10,852	10,899	11,599	2.90	3.53	3.71	3.78	4.14
Haute Normandie	10,735	11,769	10,034	10,292	10,403	3.77	4.07	3.43	3.57	3.71
Alsace	5,741	6,459	7,752	8,516	10,104	2.01	2.23	2.65	2.96	3.60
Midi–Pyrénées	5,146	5,503	5,804	6,510	8,818	1.81	1.90	1.99	2.26	3.14
Aquitaine	3,546	4,070	5,539	6,292	7,148	1.24	1.41	1.90	2.18	2.55
Basse Normandie	7,150	7,306	7,341	6,698	6,861	2.51	2.53	2.51	2.32	2.45
Nord–Pas-de-Calais	7,770	7,574	6,339	6,231	6,225	2.73	2.62	2.17	2.16	2.22
Franche Comté	6,338	6,160	5,666	6,185	5,805	2.22	2.13	1.94	2.15	2.07
Bourgogne	6,790	6,696	5,532	5,341	5,344	2.38	2.32	1.89	1.85	1.91
Languedoc–Roussillon	3,242	3,422	3,423	4,607	4,518	1.14	1.18	1.17	1.60	1.61
Lorraine	2,142	2,480	2,812	3,350	3,645	0.75	0.86	0.96	1.16	1.30
Auvergne	3,900	3,774	3,245	3,506	2,855	1.37	1.31	1.11	1.22	1.02
Poitou–Charentes	4,906	4,253	3,617	2,660	2,778	1.72	1.47	1.24	0.92	0.99
Picardie	2,352	1,877	2,346	2,369	2,025	0.83	0.65	0.80	0.82	0.72
Limousin	2,423	1,874	1,875	1,932	1,568	0.85	0.65	0.64	0.67	0.59
Champagne–Ardennes	827	960	771	1,065	1,116	0.29	0.34	0.26	0.37	0.40
TOTAL	284,933	289,004	292,233	288,140	280,375	100.00	100.00	100.00	100.00	100.00

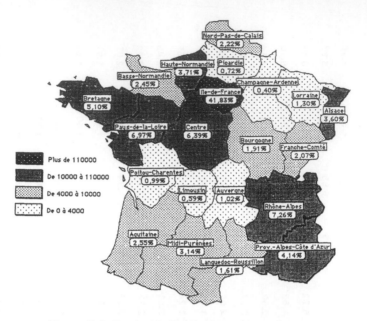

Figure 5.9 Regional distribution of employment in information industries, 1987 (*Source:* INSEE)

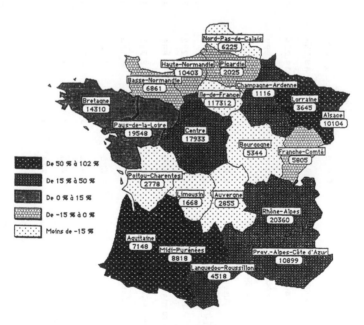

Figure 5.10 Change in information industries employment, 1975–87 (*Source:* INSEE)

electronics industry in the "désert française" resulted in the decentralization of more than 60 major operations over a period of fewer than 20 years, mainly towards the regions of the west (Table 5.10).

Table 5.10 The main information industry
relocations in France, 1957–77.

1957	LMT	Laval
1958	Thomson CSF	Brest
1960	Laboratoires de CNET	Lannion
1961	Bull	Angers
1962	Thomson CSF	Laval
	IBM	Montpellier
	CGE	Cherbourg
1963	Texas Instruments	Nice
1965	AOIP	Guingamp
1966	Motorola	Toulouse
1968	Sagem	Fougères
	Cie Int. d'Informatique	Toulouse
	Philips	Lunéville
1970	CIT	La Rochelle
1971	Ericsson	Brest
	IBM and Siemens	Bordeaux
	SAT	Dinan
	CGCT	Rennes
	AOIP	Morlaix
1972	CGCT	Boulogne sur Mer
1973	Rank Xérox	Tourcoing
	Motorola	Angers
1977	National Semiconductor	Marseille

Source: Durand (1974)

This process of decentralization, although strongly state-supported, was clearly associated with a number of other interrelated processes: (a) an increasingly tight labour market in the Paris region; (b) a significant surplus of non-qualified labour in the rural regions – mainly liberated from a rapidly restructuring peasant agriculture – and in the declining industrial regions; (c) the decentralization of the defence industry towards the south during the interwar period; and (d) the expansion of Taylorist organizational structures in the electronics industry which facilitated a functional spatial division of labour.

In short, the Taylorist rationalizations of industrial capital converged with the objectives of the state to promote decentralized development.

Moreover, the decentralization of information industries did not really entail a geographical displacement of jobs (Pelata & Veltz 1985). Indeed, the percentage of jobs actually relocated from Paris to the regions was extremely small. What actually happened was that the employment base of those sectors expanded in the provinces, and the skill and qualification structure of the labour force was spatially recomposed. The geographical displacements that did take place mainly concerned low-skilled, unqualified jobs. The total number of jobs in Paris continued to increase, but the skill composition changed completely. The number of low-skilled jobs fell dramatically while the number of top-level jobs increased rapidly. As such, the decentralization process has not affected the spatial concentration of high-level jobs in the same way as it did with low-skill categories. The main effect of the decentralization process, therefore, was a general reduction of the regional disparities in terms of the *number* of jobs, while the qualitative differences became even more pronounced. This period of spatial redeployment came to a standstill around 1975 to make way for a completely different form of territorial restructuring (Swyngedouw & Anderson 1987).

The territorial recomposition of the information industry after 1975
After 1975, the sectoral composition of the various regions in terms of type of firms and of activities, the qualification structure of the labour force, the degree of intersectoral mixing, the level of foreign investments and so on – all became crucial determinants in the process of regional recomposition of the information industry.

Three regions show a mainly mono-sectoral composition. Nord–Pas-de-Calais specializes in telecommunications, which represent 65% of total information industry employment in 1975, but this share gradually declined over the subsequent years to 42% in 1987, while the computer industry expanded. However, the region's share in the national total is extremely small, at only 2.2% in 1987. The decentralization of the tele-communications industry affected Bretagne in a most significant way. In 1978, this sector represented 85% of total information industry employment. After 1978, the importance of this sector fell relative to *l'électronique professionnelle*. The crisis in the telecommunications sector in France after 1978 – after the "plan telecommunication" of the French government drew to a close – hit Bretagne particularly hard and demonstrated forcefully the potentially disruptive effects of mono-sectoral (and state-dependent) specialization. The subsequent reconversion of the telecommunications industry (mainly towards other information-based activities) was slow and painful, and was largely dependent on the development of a local network of suppliers and subcontractors. On the whole, however, Bretagne continued to increase both its share and absolute number of jobs in information industries.

The third mono-sectoral region is Languedoc–Roussillon. Although not very important in absolute terms, its specialization is in the most dynamic information-based sector, i.e. computing. This sector accounts for 70% of total C&C employment in the region. This regional specialization stems from the successful decentralization of IBM Montpellier in 1963, which subsequently generated a host of local supplier firms as well as other, independent, information-based product innovators.

Haute Normandie is an interesting case in the sense that, until 1983, the region's C&C employment was mainly concentrated in telecommunications and passive components. In recent years, however, there was a marked restructuring of the region's C&C manufacturing base, resulting in a quite important recomposition of the sectoral mixture in the direction of *l'électronique professionnelle* and the production of active components. The latter is mainly attributable to the location of a component production and assembly branch of Thomson in Ymarle in 1987.

Of these four regions, all with a more or less distinct specialization, three have been heavily influenced by the decentralization policy in the telecommunications industry and are characterized by an above-average share of low- or unskilled workers in the composition of the industry's labour force. Moreover, the decentralized activities remain largely dependent on external (Paris-based) decision centres.

The other important geographical areas in France are the regions west of Paris, which have been affected by decentralization tendencies in important ways, and the southern regions which in recent years have shown the most dynamic employment growth. The key regions in the area west of Paris are mainly Pays de Loire and Centre, and, to a lesser extent, Basse Normandie. These regions show a highly diversified electronics and information industry, with a relatively high concentration in the production of mass-consumption electronics. In Pays de Loire and Centre, telecommunications employment falls (as in most other regions), while *l'électronique professionnelle* and informatics employment tend to rise. In Basse Normandie, there is a marked tendency towards specialization in sectors which demand a lower skill structure of the labour force, i.e. mass-consumption electronics and active components (mainly semiconductors).

Finally, the southern regions, i.e. Rhône–Alpes, Provence–Côtes d'Azur

(PACA), Midi–Pyrénées and Aquitaine, are the most dynamic and diversified regional economies in terms of C&C activities, although Midi–Pyrénées shows a somewhat more pronounced specialization in computers and semiconductors. As Beckouche (1987) notes, some of these regions already contained an important electronics industry in the 1950s, on which decentralized Parisian industries were grafted from the 1960s onwards. Moreover, as will be discussed below, the quality of the activities in the southern regions is decidedly higher than of those in the other provincial regions.

In Rhône–Alpes, the development is centred mainly around the micro-electronic poles of Grenoble and Lyon. These areas benefit from many technology centres and important science-based educational facilities.

The growth of the electronics industry in PACA followed the arrival of both IBM and Texas Instruments in the region in 1963, although the interwar decentralization of defence industries had already created an innovation seedbed. The creation of Sophia–Antipolis in the late 1970s, which became a significant science and research area (Thomson, DEC) (Perrin 1987), further consolidated the rise of PACA as a centre of C&C excellence. As we shall document below, the computer industry (*l'informatique*) accounts for 22% of total C&C employment in the region in 1987. The electronics-based growth of Midi-Pyrénées stems from decentralized (international) branches in conjunction with state intervention (in defence) and sustained by nationalized aeronautics industries. This complex network provided an incubator for the growth and development of C&C industries.

Similar observations can be made for Aquitaine. The aerospace industry followed the establishment of branches of IBM at Canejean, and at Bordeaux, Thomson–CSF, Siemens and SAT. Since the 1980s, more mass-consumer orientated activities came into the region, mainly of foreign origin (Sony, Pioneer).

The Île-de-France region shows a remarkable and interesting evolution towards a higher degree of specialization. In 1987, two sectors dominate the C&C scene, *l'électronique professionnelle* and information technologies (*l'informatique*). In 1975, the regional C&C structure was much more diversified. A swing away from diversification towards specialization is, consequently, one of the most characteristic tendencies in the Paris region. This narrowing down of the range of C&C activities is the result of a global job loss in C&C industries, with the exception of computers and l'électronique professionnelle, although, since the early 1980s, the latter sector has been losing jobs in the region. The most dynamic sectors reinforce their position in Île de France, while consumer electronics and telecommunications are fading away rapidly.

In conclusion, the regions west of Paris were most heavily affected by decentralization processes and they inherited a sectoral specialization based upon telecommunications. However, there is a concentration of less dynamic and non-qualified activities, even if computers and *l'électronique professionnelle* occupy a significant position in those regions. They underwent massive restructuring during the 1980s, resulting from the contraction and crisis of the telecommunications sector. In most cases, the job losses in C&C sectors in those regions were in telecommunications.

The southern regions have a much more dynamic and diversified C&C structure. They succeeded in creating industrial synergies and in developing an integrated C&C network which benefits this zone considerably. This structural advantage of the south over the west is reflected in the employment qualification structure (see below).

Evolution of number and size of firms
The French regions are also differentiated in terms of the size of firms (see Table 5.11). The available data (from ASSEDIC) need to be carefully interpreted. First, it is impossible to distinguish the internal dynamics of small and large firms. Secondly, the available information does not allow differentiation between newly created firms and branches established by large firms. Given these remarks, the average size of C&C firms in France in 1987 was 45, a fall from 92 in 1977. Nevertheless, there are marked inter-regional differences. The size of firms is above average in the western regions and below average in the south. Île-de-France has an intermediate position (average of 50). Figure 5.11 highlights these structural regional differences. This pattern also confirms the conclusions made at the end of the previous section. Over the 1977–87 period in France, the number of firms in C&C sectors doubled, while the number of jobs increased only slightly (Fig. 5.12). It is again possible to distinguish between the three core zones of C&C activity. In the western regions employment declined while the number of firms grew; in the south these two variables increased and converged. The number of jobs fell in Île-de-France, which accounts for 25% of the total number of firms. These tendencies confirm the evidence presented by Swyngedouw & Anderson (1987), who noted a marked regional discrepancy in the relationship between growth in employment and the number of firms in high-technology sectors. The positive relationship between these two variables at the national level does not translate into a uniform regional picture.

107

Table 5.11 Change in the number and size of establishments by region, 1977–87.

Regions	Employment			Change in employment (%)			Number of establishments		
	1977	1981	1987	1977–81	1981–87	1977–87	1977	1981	1987
France	311,016	313,695	318,144	0.86	1.42	2.29	3,399	4,643	7,045
Ile-de-France	153,256	149,205	138,221	-2.64	-7.36	-9.81	1,613	1,977	2,776
Champagne–Ardennes	1,105	1,207	1,402	9.23	16.16	26.88	37	47	76
Picardie	2,119	2,543	2,511	20.01	-1.26	18.50	48	85	114
Haute Normandie	12,907	10,783	10,910	-16.46	1.18	-15.47	94	111	155
Centre	19,015	18,040	18,211	-5.13	0.95	-4.23	137	170	253
Basse Normandie	7,424	6,759	6,575	-8.96	-2.72	-11.44	52	60	90
Bourgogne	7,234	5,656	6,250	-21.81	10.50	-13.60	61	85	116
Nor–Pas-de-Calais	8,106	7,041	6,674	-13.14	-5.21	-17.67	99	143	214
Lorraine	2,625	3,632	4,349	38.36	19.74	65.68	75	122	184
Alsace	5,297	5,873	8,552	10.87	45.62	61.45	66	102	151
Franche Comté	6,278	5,982	5,357	-4.71	-10.45	-14.67	48	55	88
Pays de Loire	20,535	21,399	19,799	4.21	-7.48	-3.58	131	197	308
Bretagne	11,978	14,169	15,092	18.29	6.51	26.00	88	152	245
Poitou Charentes	4,455	3,934	3,480	-11.69	-11.54	-21.89	38	56	85
Aquitaine	4,375	6,258	8,211	43.04	31.21	87.68	109	157	262
Midi–Pyrénées	6,014	7,306	10,010	21.48	37.01	66.44	106	181	308
Limousin	1,863	2,069	1,814	11.06	-12.32	-2.63	28	39	48
Rhône–Alpes	18,175	21,967	25,317	20.86	15.25	39.30	305	439	759
Auvergne	3,951	3,657	4,993	-7.44	36.53	26.37	41	62	98
Languedoc–Roussillon	4,252	4,036	5,731	-5.08	42.00	34.78	45	83	199
PACA	10,035	11,826	14,385	17.85	21.64	43.35	177	302	493
Corse	0	71	142	–	100.00	–	0	7	14

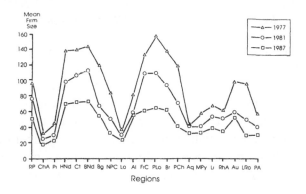

Figure 5.11 Change in the mean firm size, by region, 1977–87
(Source: ASSEDIC)

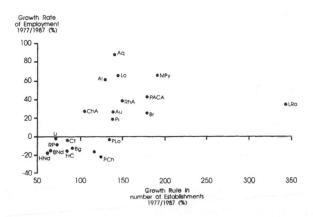

Figure 5.12 Regions by growth of employment and number of firms, 1977–87 *(Source:* ASSEDIC)

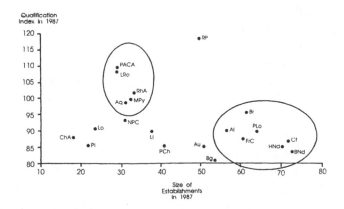

Figure 5.13 Regions by qualifications and size of firm, 1987
(Source: ASSEDIC)

109

Finally, Figure 5.13 reveals an important relationship between firm size and qualification structure of the labour force. While the Paris region is distinctive, there is marked difference between the western and southern regions. The western zone (as well as the eastern regions Alsace and Franche Comté) clusters around a profile made up of large firms with a predominantly low labour qualification structure. The southern regions, on the contrary, show a completely different pattern, characterized by relatively small firms and a high qualification structure of the labour force.

Regional pattern and evolution of information-equipment production and services in France

Figures 5.14–28 detail the regional evolution of selected C&C sectors over the 1981–87 period. The computer industry and office-equipment employment (NAP sectors 2701 and 2702), presented in Figures 5.14–16 is concentrated predominantly in the Paris region, which, of course, provides a major consumer market for these final consumer products. Indeed, the location dynamics of this kind of production are influenced by market conditions, putting a premium on agglomeration economies and close contacts with customers. The southeastern regions, as well as the belt around Île de France and Bretagne, consolidated or improved their position *vis-à-vis* other regions.

The electronic components industry (NAP 2915 and 2916), presented in Figures 5.17–19, can be characterized as vertically integrated, i.e. oriented towards mass-assembly production which tends to promote the location of production facilities in lower-cost areas. This industry experienced a large net growth in employment which was concentrated partly in Paris, but with additional growth in the south and west, where labour costs are lower. The dominance of Paris is much less pronounced and the "peripheral" regions are doing reasonably well.

The large employment losses in the telecommunications industry (NAP 2911) were felt primarily in the Paris region, with smaller losses in Normandy, Nord–Pas-de-Calais and the Loire. Bretagne, where there was a significant concentration of the industry at the beginning of the period, actually experienced a small gain in employment and saw its share in national employment rise from 14.9% to 16.6% over the 1981–87 period, while Paris fell from 41.4% to 33.1% (see Figs 5.20–22). Alsace and the south-east also improved their relative positions over this period.

Figures 5.23–25 document the regional distribution of *l'électronique professionnelle* (NAP sectors 2912, 2913, 2914). A similar regional pattern to that observed for the other sectors is evident for these sectors. In

110

addition, the south-west accounts for a relatively large share of these activities. This is associated with the high percentage of employment and investment in the aerospace and defence-related industries in these regions. Indeed, about 60% of the output of these sectors is procured by the defence industry, a key factor in the regional coincidence of these activities (see Swyngedouw 1988).

Finally, information and telecommunications services are concentrated in the Paris region, but their share fell during the 1980s (Figs 5.26–28). Regions with growth in C&C activities or consolidating their position during the 1980s also did so for telecommunications and information services. The regional correlation between C&C manufacturing and C&C services is quite clear and it confirms the demand-led development of these services.

Qualifications structure of the labour force in C&C industries in the UK and France

The United Kingdom

It is extremely difficult to compile reliable data on the skills and qualifications profile in C&C industries in the UK. Most observations made here are compiled from the UK Engineering Industry Training Board's (EITB) profile of British engineering (1989). The labour force in engineering industries fell sharply by one million between 1978 and 1988, the decline in the early 1980s being particularly dramatic. Employment has declined most in relatively low-skilled occupations such as operators, clerical staff, supervisors and craft workers. By contrast, employment of professional engineers, scientists and technologists has risen by 55% since 1978. Women represent only 20% of the engineering work-force, the vast majority of them employed as clerical staff or operators.

Data from the Department of Employment Census indicates that, for the four core C&C Standard Industrial Classification classes (SICs), female employment accounted for about a quarter of the total, and that part-time employment was significant only in telecommunications equipment (11% of total in 1984), with no significant changes in those proportions.

The EITB provides a much richer source of information on the skill composition of regional work-forces in electronic data-processing and telecommunications equipment (Tables 5.12 & 13).

The regional statistics on changes in the qualifications structure of C&C industries in the UK display a number of significant facets: (a) a generalized shift towards greater proportions of professional workers, and

111

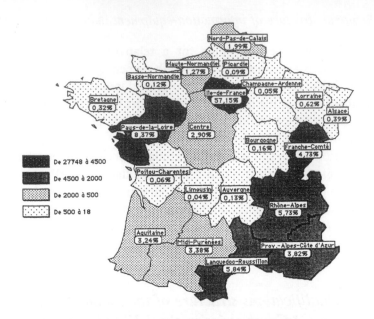

Figure 5.14 Regional distribution of employment in informatics industries, 1981 NAP sectors 2701 (data-processing equipment), 2702 (office machinery). (*Source:* INSEE, ESE 1983)

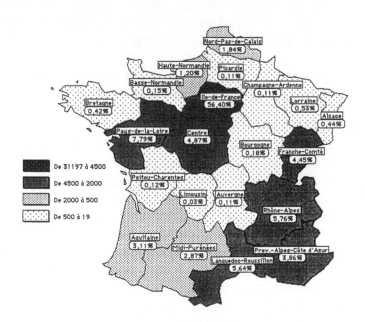

Figure 5.15 Regional distribution of employment in informatics industries, 1983 NAP sectors 2701 (data-processing equipment), 2702 (office machinery). (*Source:* INSEE, ESE 1983)

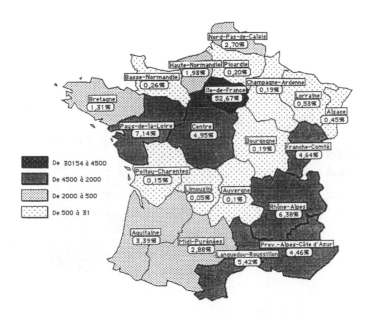

Figure 5.16 Regional distribution of employment in informatics industries, 1987 NAP sectors 2701 (data-processing equipment), 2702 (office machinery). (*Source:* INSEE, ESE 1983)

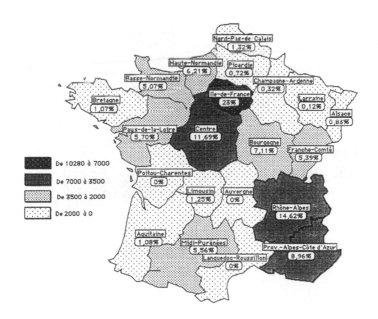

Figure 5.17 Regional distribution of employment in electronic components, 1981 NAP sectors 2915 (passive components), 2916 (semiconductors). (*Source:* INSEE, ESE 1981)

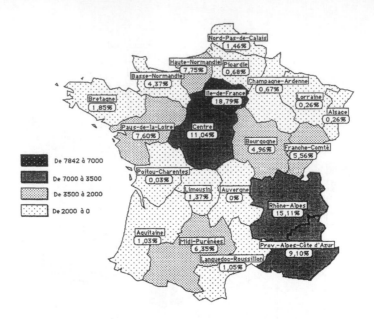

Figure 5.18 **Regional distribution of employment in electronic components, 1983** NAP sectors 2915 (passive components), 2916 (semiconductors). (*Source:* INSEE, ESE 1983)

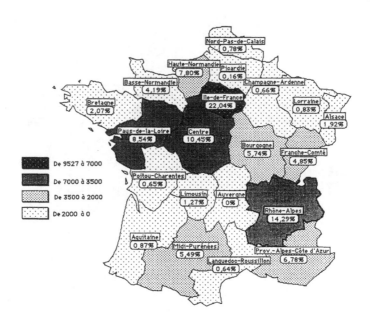

Figure 5.19 **Regional distribution of employment in electronic components, 1987** NAP sectors 2915 (passive components), 2916 (semiconductors). (*Source:* INSEE, ESE 1987)

114

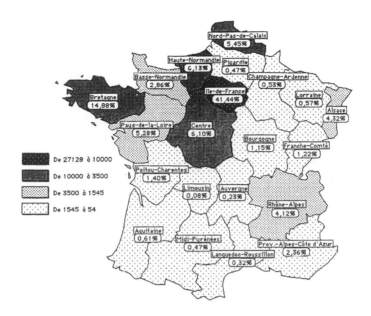

Figure 5.20 Regional distribution of employment in telecommunications equipment, 1981 NAP sector 2911 (telegraphic
and telephone equipment). (*Source:* INSEE, ESE 1981)

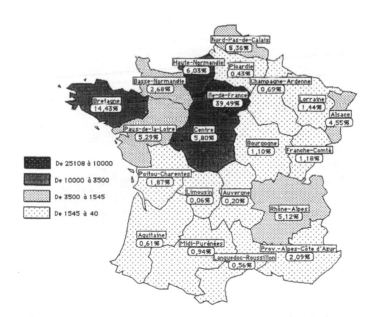

Figure 5.21 Regional distribution of employment in telecommunications equipment, 1983 NAP sector 2911 (telegraphic
and telephone equipment). (*Source:* INSEE, ESE 1983)

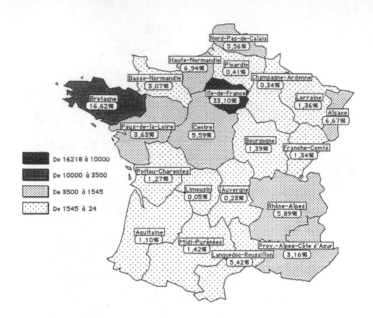

Figure 5.22 **Regional distribution of employment in telecommunications equipment, 1987** NAP sector 2911 (telegraphic
and telephone equipment). (*Source:* INSEE, ESE 1987)

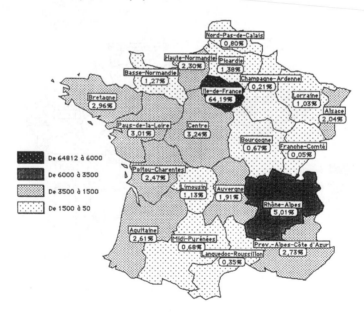

Figure 5.23 **Regional distribution of employment in scientific
instruments, 1981** NAP sector 2912 (medical electronics),
2913 (measurement & control instruments), 2914 (scientific
and radio electronics). (*Source:* INSEE, ESE 1981)

116

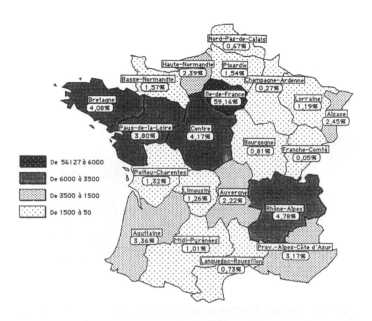

**Figure 5.24 Regional distribution of employment in scientific
instruments, 1983** NAP sector 2912 (medical electronics),
2913 (measurement & control instruments), 2914 (scientific
and radio electronics). (*Source:* INSEE, ESE 1983)

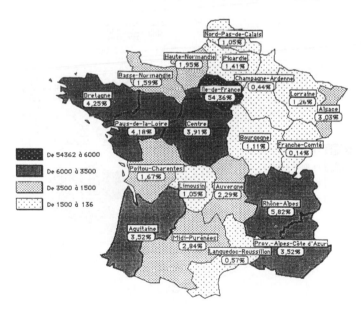

**Figure 5.25 Regional distribution of employment in scientific
instruments, 1987** NAP sector 2912 (medical electronics),
2913 (measurement & control instruments), 2914 (scientific
and radio electronics). (*Source:* INSEE, ESE 1987)

117

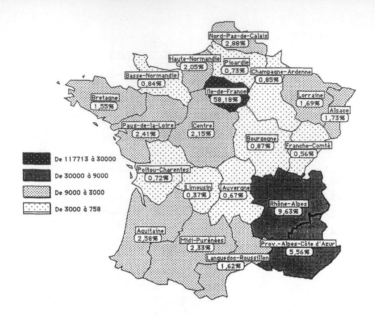

Figure 5.26 Regional distribution of employment in informatics and telecommunications services, 1981 NAP sectors 75 (telecommunications & post), 7701 (technical studies), 7702 (economic & social studies), 7703 (informatics studies), 7704 (informatics applications). (*Source:* UNEDIC 1981)

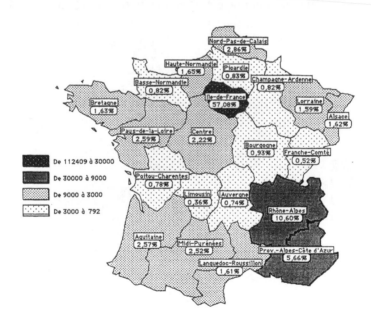

Figure 5.27 Regional distribution of employment in informatics and telecommunications services, 1983 NAP sectors 75 (telecommunications & post), 7701 (technical studies), 7702 (economic & social studies), 7703 (informatics studies), 7704 (informatics applications). (*Source:* UNEDIC 1983)

118

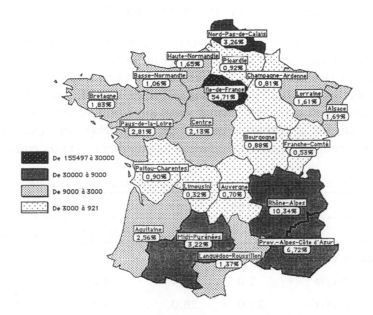

Figure 5.28 Regional distribution of employment in informatics and telecommunications services, 1987 NAP sectors 75 (telecommunications & post), 7701 (technical studies), 7702 (economic & social studies), 7703 (informatics studies), 7704 (informatics applications). (*Source:* UNEDIC 1987)

Table 5.12 Electronic data-processing: percentage of total workforce.

	1983		1989	
	Professionals	Operators	Professionals	Operators
South East	26.8	14.6	31.9	14.0
East Anglia	28.2	11.7	23.0	17.9
South West	21.0	25.0	29.5	15.0
West Midlands	24.2	27.0	27.5	24.9
East Midlands	27.1	21.1	22.3	37.2
North West	23.8	23.8	38.9	16.3
Yorkshire & Humberside	19.5	14.2	23.1	16.6
North	9.0	43.6	23.1	27.8
Wales	14.0	51.1	17.6	54.5
Scotland	25.7	31.8	22.9	37.2
Great Britain	24.9	21.5	29.5	21.9

Source: Engineering Industry Training Board statistics
Note: "Professionals" comprises managers, engineers, scientists and technologists

119

relative decline in the proportion of operators; (b) the continued importance of the South East as a location for professional employment, and a continuing concentration of operator-grade workers in regions such as Wales, Scotland and northern England; (c) a rise in the relative importance of the South West in terms of management and scientific employment; (d) a wider variation in regional shares of professional/operator employment within telecommunications equipment as opposed to computer equipment.

Table 5.13 Telecommunications equipment: percentage of total workforce.

	1983		1989	
	Professionals	Operators	Professionals	Operators
South East	18.3	29.6	22.2	27.3
East Anglia	2.9	59.4	4.0	35.1
South West	21.0	25.0	43.2	17.2
West Midlands	16.8	41.0	33.5	28.7
East Midlands	12.0	39.3	25.5	26.7
North West	8.1	47.8	38.6	33.6
Yorkshire & Humberside	4.0	52.0	11.3	43.3
North	4.6	64.1	13.2	60.1
Wales	8.3	64.6	8.8	62.4
Scotland	3.3	67.2	11.7	59.3
Great Britain	13.0	43.5	22.4	34.6

Source: Engineering Industry Training Board statistics
Note: "Professionals" comprises managers, engineers, scientists and technologists

France

In the French C&C industries between 1975 and 1987, there was: (a) an overall decline in the total number of jobs, at least for most sectors; (b) an increasing substitution of manual workers by engineers and technicians; (c) an improvement in the overall qualifications structure of the labour force, resulting from a dramatic fall in the number of non-qualified manual workers. However, the regional profile of the qualifications structure of the labour force may reveal important tendencies, in terms of both the overall restructuring of the industry and of its regional dynamics (Fig. 5.29).

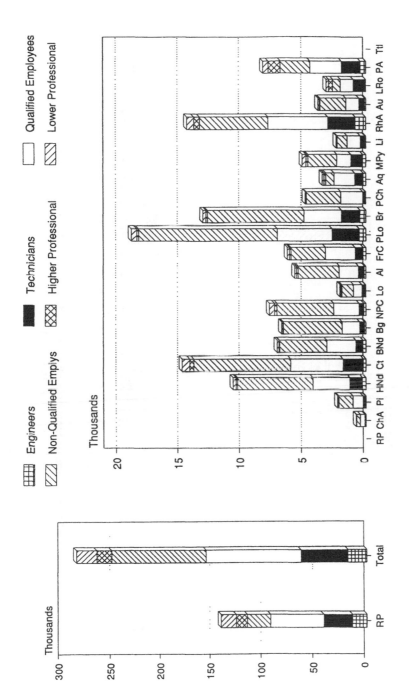

Figure 5.29a Regional occupational change in information industries, 1975 (%)
(*Source:* INSEE)

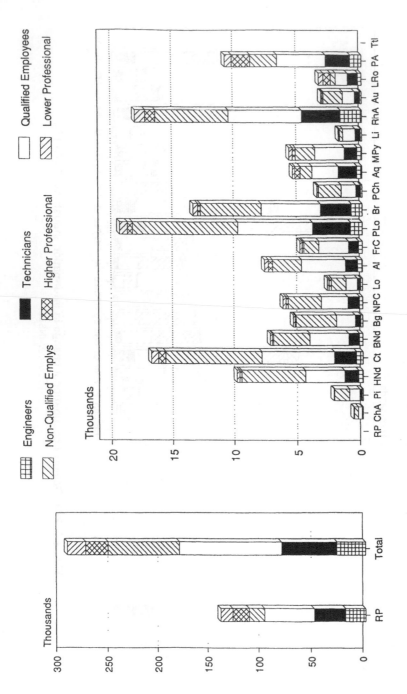

Figure 5.29b Regional occupational change in information industries, 1981 (%)
(*Source:* INSEE)

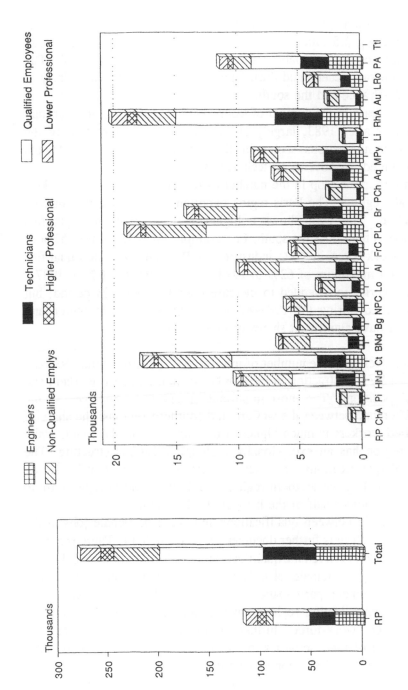

Figure 5.29c Regional occupational change in information industries, 1987 (%)
(*Source:* INSEE)

These data show:

(a)   An important growth in the number of engineers in all regions (increasing 2.5 times for France as a whole), but reaching a peak of 60% of the total labour force in the Paris region in 1987. The growth rate accelerated during the 1982–87 period, particularly in Île-de-France and the south.

(b)   An increase in the number of technicians, despite a fall in their total number after 1983, largely attributable to the contraction of the telecommunications sector in Île-de-France, which nevertheless accounted for 50% of all technicians.

(c)   An important drop in the number of non-qualified manual workers in all regions. The Paris region experienced the most pronounced decline. The west shows a higher percentage of non-qualified workers than does the south. For example, in 1975, almost 57% of the work-force in C&C industries in Bretagne was non-qualified (with a maximum of 62% in 1978), but this fell to 20% in 1987.

(d)   Managerial jobs tended to increase over the 1975–83 period, but decreased during the 1983–87 period. Only the Paris region and PACA have significant shares of managerial jobs.

In short, the decline in the number of jobs in C&C industries after 1981 can be explained by a contraction in the number of non-qualified workers and managerial positions. The most important observation, however, is the marked difference between the western and southern regions: the share of non-qualified workers is much higher in the west than in the south, while the Paris region has an exceptionally high qualifications structure. The hierarchy of qualifications structures, therefore, is dominated by Île de France, followed by the southern regions, while the western regions are to be found in the lower half of the list (see Table 5.14).

The relationship between qualifications and share of female jobs in the employment structure is further documented in Table 5.14. There is a clear inverse correlation between the qualifications composition of the work-force and the percentage of female jobs. The southern regions and Paris show a relatively low female participation rate compared with the west. The traditional decentralization regions have an above-average percentage of women in their work-force, further strengthening the thesis that the decentralization of C&C activities to the western regions was based on a very specific functional division of labour.

The analysis of the regional qualifications structure of the labour force

Table 5.14 Classification of regions (1987): qualifications index (QI) and percentage of women in the labour force.

Region	QI	Total women	% women
Ile-de-France	118.4	117,312	29.0
PACA	109.3	11,599	30.5
Languedoc–Roussillon	108.0	4,518	20.2
Rhône–Alpes	101.4	20,360	35.0
Midi–Pyrénées	99.6	8,818	36.1
Aquitaine	98.3	7,148	33.6
Bretagne	95.5	14,310	44.9
Nord–Pas-de-Calais	92.8	6,225	38.0
Lorraine	90.5	3,645	40.7
Alsace	89.9	10,104	38.0
Pays de la Loire	89.7	19,548	46.5
Limoges	89.0	1,668	25.5
Champagne–Ardennes	87.5	1,116	36.8
Franche Comté	87.2	5,805	44.2
Centre	86.9	17,933	45.7
Picardie	85.3	2,025	44.0
Haute Normandie	85.1	10,403	48.9
Poitou Charente	84.9	2,778	52.1
Auvergne	84.6	2,855	44.1
Basse Normandie	83.6	6,861	54.4
Bourgogne	80.5	5,344	53.2

Source: INSEE Lemattre, M.

converges with the analysis of employment and firm dynamics. The southern regions combine sectoral diversity with a positive employment dynamic and a high-labour force qualifications pattern. The western regions suffer from the combined effects of sectoral specialization in less dynamic industries and a rather low-skill, female-based, job composition.

Conclusions

Clearly, the key C&C industries in the UK are in a dynamic phase of restructuring and development. The South East of England is the dominant employment pole. However, within that regional pre-eminence London has become a relatively insignificant centre of employment. For example, in 1981, in London more were employed in electronic data-processing manufacture than in Scotland, but fewer in 1987. London nevertheless remains an important centre of employment in computer and telecommunications services, as does the South East, which moreover has increased its share of employment while London's computer and telecommunications services share has declined relative to that of the rest of Britain. In telecommunications-equipment manufacture London has lost one-third of its jobs and by 1987 employed fewer than in Wales and four other regions. Proportionally, many of the regions of outer Britain have increased their share of C&C manufacturing employment.

The other general feature of the key industries has been the transformation of the occupational structure. All regions display an increase in the proportion of professionals between 1983 and 1989. In every case this growth has been at the expense of the operator work-force. Some regions, such as the South West and Midlands (East and West) have doubled their professional work-forces in telecommunications equipment industries. The larger increases in professional employment in electronic data-processing have occurred in non-metropolitan Britain.

The regional dynamics and structure of French C&C industries reveal some distinct characteristics. In general, the Paris region is extremely dominant, in both quantitative and qualitative terms; the relative decline of Paris in the manufacturing sectors is largely compensated for by its dominance in information and C&C services. The regions west of Paris were most affected by the decentralization policy of the 1960s and 1970s, particularly in the telecommunications sector. Those regions consequently inherited a predominantly mono-sectoral specialization, an employment basis built around a low-skilled, often female, labour force. They were largely dependent on the Paris-based decision centres. These factors made them highly sensitive to sectoral crises, resulting in major restructuring problems during the telecommunications crisis of the early 1980s. In contrast, the southern regions, whose take-off in C&C activities dates back to approximately the same period (1960s and 1970s), benefited in a way quite different from the decentralization process. Their diversified sectoral composition, their highly qualified labour force, the small firm size and

126

their relative independence from Paris resulted in a highly dynamic regional economy, based on high technology and advanced services.

The rise of the south poses a number of interesting questions about the dynamics of regional growth. The decentralization rationale towards the south during the 1960s was exactly the same as the motivation behind decentralization to the west, i.e. abundance of cheap non-qualified labour during a period of rapid rationalization and Fordist transformation of the production process. However, for French companies, the southern regions originally suffered from the major drawback of being far from Paris. This explains – at least partially – the preference on the part of French firms to decentralize towards the West. For example, Pelata & Veltz (1985) mention that the top management of Bretagne's key telecommunications plants filled a commuter plane to Paris every day. This proximity to Paris did not hold for foreign (mainly US) firms. These, together with the decentralized aerospace industry, formed the basis for C&C industry development in the south. For example, IBM's location in Montpellier benefited both from the incentives coming from DATAR (*Délégation à l'Aménagement du Territoire et à l'Action Régionale*) and from a large, available pool of workers in the south, as well as from the proximity of important educational centres. As Pottier & Touati (1982) argue, American plants locating in France concentrated in a few large southern metropolitan areas. For example, Nice (Texas Instruments) and Toulouse (Motorola), offered the classic labour-force advantages but also a high-quality environment for American professionals, excellent international communications lines and outstanding educational and research facilities. This dominance of foreign firms in the south reduced the area's dependence on Paris, while the labour-force qualifications and research potential of those regions improved steadily. IBM and Texas Instruments established research centres in addition to their production facilities. These elements produced a highly dynamic C&C environment in the south which was not created in the west.

A rather similar explanation can be offered for the spatial changes found in the UK. Older equipment manufacturers moved away from London, through either closures or relocations away from the area, particularly in telecommunications equipment. New computing firms moved in from the US, Japan and Europe to the outer South East, Scotland and the South West (Morgan & Sayer 1988). The qualifications structure of these "transplants" was more varied, with high proportions of professional workers. Elsewhere, growth occurred in regions characterized by declining heavy industry, but in these regions a greater proportion of jobs tended to be in the operator grades. However, even in such regions, the relative proportions of professional-grade workers were increasing towards the end of the 1980s.

In sum, regional employment and occupational dynamics in C&C industries depends crucially on the place each region takes in the division of labour in its various forms, i.e. sectoral, level of dependency, functional, and so on (Valeyre 1982). The location dynamics of C&C industries have clearly contributed, at least after 1975, to a new model of development and spatial behaviour and, consequently, a new qualitative regional hierarchy.

CHAPTER SIX

The computer hardware industry in the 1980s: technological change, competition and structural change

Peter Wells & Philip Cooke

The 1980s have seen dramatic developments in the computer hardware industry, with rapid rates of technological change, reduced product life cycles and the increased penetration of computer applications into the economy, decisively altering the conditions for competitive success. In the process the simple hierarchy of large, medium and small mainframe products has been increasingly blurred, both by the emergence of new hardware configurations such as super-minis, workstations and laptops, and by the development of new value-added services markets, especially in computer networks. Additionally the peripherals element of computer hardware has taken on a greater significance. Clearly, any account of such a large, diverse and changing sector must be partial and selective. As such, this chapter identifies major trends and changes in the industry at a global level, before a closer analysis of the UK and France is presented which incorporates our sample firms.

The first section is concerned with the segmentation of the computer hardware industry at the global level. Quantum improvements in the performance of underlying technologies together with the development of more powerful software have combined with major changes in the uses to which computers are put. In the process the domination of IBM has been challenged by a number of dynamic young firms which have successfully exploited the market niches created by product and service segmentation. Moreover, the pre-eminence of US firms and the US market on the global scale is also under attack (Flamm 1987, 1988). More importantly still, the drive by major users (both corporate and state) for networked computers is forcing the hardware firms into the services market – notably in terms of systems integration – for which they may not yet be properly prepared.

Europe has provided one of the major growth markets of the 1980s, but, as the second section shows, the European computer hardware industry has been less successful than US firms in reaping the benefits of this growth. The second section concentrates on the hardware industries of the UK and France. In both cases US firms constitute a significant proportion of the industrial base; indeed, in many cases the US firms already have an integrated European structure of research, production and marketing which European firms cannot match. However, the policy perspectives of the state in the UK and France have been very different in the 1980s (Flamm 1987), which in turn has led to stark differences between the two main domestic manufacturers, ICL and Groupe Bull. While ICL has counted the costs of commercial independence and freedom from overt state intervention, especially in terms of large scale rationalization, it could be argued the firm is better fitted to compete on world markets than the "national champion" Groupe Bull, especially since ICL's acquisition by the Japanese giant, Fujitsu, in 1990. Even without the emergence of a new Sun Microsystems or Compaq to challenge US dominance, or the arrival of Japanese (Fujitsu–ICL excepted) computer hardware firms, several of the European niche firms such as Nixdorf and Norsk Data have recently collapsed in the face of growing competition.

Having discussed technological, competitive and market changes, the final two sections are concerned with an analysis of how firms have responded to (and helped initiate) the changes noted. For ease of exposition the analysis is divided into two sections concerning organizational dynamics: the first internal, the second external.

Internal organizational dynamics refers to the structure and composition of work processes within the firm, for all types of employees. The 1980s have witnessed huge upheavals within very large corporations as they have attempted to create organizational structures which more adequately address changing production and market requirements. ICL is an especially interesting example both of the application of new organizational principles and of the difficulty of enforcing that application in the face of its convergent (STC in telecommunications) corporate culture. Nonetheless, all computer hardware firms are facing the same difficulties in terms of reconciling declining product life-cycles and the demand for more flexible production to suit particular market applications.

External organizational dynamics refers to the relations between the firms and the "outside world", chiefly but not exclusively in terms of other corporations and various state bodies. Here the commodification of the computer hardware market, together with the need to focus corporate resources on core competences and technologies, is driving the trend towards greater vertical disintegration – hardware firms are increasingly assemblers rather than producers as such. In this process, key suppliers both up stream (components) and down stream (software) are taking on a greater significance, chiefly in terms of getting new product–service configurations onto the market quickly enough, and alliances are frequently the vehicle for achieving the required functional integration. On the other hand, the pressure of intense competition and the need to spread overheads over wider geographical markets has led to some rationalization of the computer hardware industry, especially the US mainframe firms. It has also led to what may be termed quasi-horizontal alliances for market penetration. Moreover, the relative failure of ownership strategies to secure for computer hardware firms the communications skills they felt they required has made alliances of strategic importance for some firms.

Global industry in the 1980s: technology, segmentation and competition in world markets

As noted above, the 1980s have been volatile and sometimes traumatic years for the global computer hardware industry. The apparently overwhelming dominance of IBM, while still intact, has been significantly eroded by the emergence of new hardware suppliers within the industry and by competitive threats from outside. Concurrently, some of the most dynamic and fastest-growing firms of the early 1980s have themselves been caught by the pace and direction of changes in the industry – a perennial problem for the niche firm. Three major trends are discussed in detail:

(a)   Major improvements in the price:performance ratio, especially from smaller machines, due principally to advances in the underlying technologies of computer hardware.

(b)   A shift away from stand-alone data-processing applications based on hierarchical computer systems grouped around corporate mainframes towards the interlinking of machines via networks.

(c)   The consequent threat to proprietary operating systems (the software which provides the basic ground-rules for computer operations) and the rise of so-called "open system" architectures based on common operating principles.

Overlaid on these three trends is the question of convergence between communications and computers. While many analysts noted convergence in the 1980s, especially those who focused on technologies (Arnold & Guy 1986), and while many firms in each sector attempted to develop complementary capabilities to capture new markets expected to appear around convergence – this was for example a key part of the break-up of AT&T (see Crandall & Flamm 1989) – it has become clear that even in purely technical terms convergence has more than one form. Given the major differences between the two sectors, it is hardly surprising that convergence has been so problematic even for firms of the stature of IBM, Ericsson and AT&T.

The price:performance ratio and the emergence of the personal computer
Significant advances in the production of computer components, especially in memory chips, have given rise to smaller machines, at lower cost, capable of performing tasks previously only met by larger machines. Moreover, the availability of comparatively cheap memory capacity has in turn supported the development of memory-intensive applications – most notably computer graphics – which has made computer use much more accessible to non-technical users.

In particular the development of successive generations of memory chips (D-RAMS) from 16kbytes in the late 1970s to 4Mbytes in the late 1980s, and with 64Mbytes a distinct possibility in the 1990s, has meant greatly increased power can be packed into smaller machines. About 76% of the world production of D-RAMS comes from Japan, a strategic weakness which prompted elements of the US hardware industry to create *US Memories*, conceived as an industry-wide alliance to produce an indigenous source. Despite the backing of IBM, DEC and Hewlett–Packard, as well as four major chip manufacturers (Intel, AMD, National Semiconductor and LSI Logic), the consortium could not attract sufficient support. Absent from the consortium were such firms as Sun Microsystems, Compaq and Apple.

Whole new products and market applications have been developed, and new companies have emerged, based around these new technologies. Mini- and midi-computers started to challenge the dominance of large mainframes in the 1970s. Especially important here was DEC with its unified VMS operating system common to the whole range of machines, an important innovation which meant that applications software could be run off different machines and which provided the structure for networking computers in a distributed system. In the 1980s the phenomenal growth of the personal computer (PC) market (and, increasingly, laptop or portable computers) has in turn challenged both the traditional mainframe companies and the midi/mini-computer specialists. Table 6.1 shows the impact of these changes in terms of shipments by US manufacturers.

Table 6.1 Worldwide shipments by US manufacturers; percentage share of product groups.

	1975	1980	1985
Desktop	0.0	6.0	20.0
Office processors	4.0	6.0	10.0
Small business	3.5	11.0	13.0
Minis	9.5	17.0	21.0
Mainframes	83.0	60.0	36.0
Total market (billions)	$12.83	$28.67	$63.31

Source: from IDC cited in Jowett & Rothwell (1986).

While the output has risen five-fold in dollar terms over the period 1975 to 1985, the decline in importance of mainframes is evident, from over 80% of output to just 36%. The most significant growth is in desktop machines (broadly speaking, personal computers), particularly in machines towards the bottom of the price range. The mainframe market is still substantial and growing, but the markets for smaller machines and for networking hardware and software are growing much faster.

Table 6.2 shows these market trends in terms of the European installed base of machines over the period 1986–92, where again the increasing significance of smaller machines is evident. Generally the European market is considered to lag some two or three years behind the USA. In the USA, sales of PCs grew rather less in 1989 than in previous years, while Europe is still experiencing the major growth phase.

The installed base of business and professional PCs is expected to quadruple between 1986 and 1992 to reach over 18 million units, and that for scientific & technical PCs (including workstations) to triple to 2.2 million units. By comparison the market for large scale systems is expected to grow by only 30%. As the output and market for smaller machines has developed, so it has also become highly competitive. Margins on hardware have been falling and product life cycles are very short. The large mainframe firms have found it extremely difficult to compete in this market – witness for example DEC's disastrous foray into PCs with its Rainbow

Table 6.2 Western Europe installed-base forecasts, 1986–93 (units by processor size class).

	1986	1987	1988	1989	1990	1991	1992
Business/professional PCs	4,715,420	6,366,455	8,596,676	10,662,747	12,760,395	14,709,305	16,449,525
Scientific/technical PCs	633,490	764,028	1,007,804	1,225,363	1,459,954	1,700,236	1,979,630
Small-scale systems	950,000	1,056,500	1,176,500	1,300,100	1,442,000	1,593,000	1,749,500
Medium-scale systems	85,950	98,800	107,080	115,180	121,850	128,800	134,030
Large-scale systems	6,226	6,866	7,115	7,395	7,595	7,788	7,990
TOTAL	6,391,086	8,292,649	10,895,175	13,310,785	15,791,794	18,139,129	20,320,675

Source: IDC market survey

range – which is part of the reason their considerable corporate resources have been directed towards high-value-added work such as customized systems integration.

The shift from stand-alone data processing

For many years, the main market for IBM and similar mainframe firms was the corporate data-processing department or large centralized users in the state sector. Often relatively autonomous from other corporate divisions, the data-processing department would be concerned with large-scale information handling in applications such as payrolls and accounts. Large firms tended to centralize such applications both functionally and spatially around the corporate mainframes.

Elsewhere within the corporate user-firms, other computers have gradually been applied, often in a piecemeal and unorganized fashion. The arrival of relatively low-cost desktop or personal computers meant that corporate data-processing departments could be bypassed. In the late 1980s, user-firms increasingly sought to integrate their diverse computer systems into interactive networks and, ultimately, to match information-systems strategy to business strategy as a whole. That is to say, the purchase of computer systems has become a strategic issue for many user-firms, a competitive weapon. By selling themselves as systems integrators, computer hardware firms will be able to sell more equipment, but also be able to reap the profits from the increased services content of their work. Indeed, the hardware firms have little choice but to develop systems-integration skills as the market and margins on stand-alone mainframes diminish. As the chapter on services makes clear, the computer hardware firms are not alone in seeking to develop systems-integration capabilities. Indeed, their relative lack of expertise in other firms' businesses compared to, say consultancy firms such as Arthur Anderson and Price Waterhouse, together with their necessarily biased view of which hardware–software combinations best suit the user-firms' needs, puts the computer hardware firms at a considerable competitive disadvantage in these markets.

The threat to proprietary systems

Closely linked to the issue of networking computers is the question of operating systems (Table 6.3). Some features stand out in this table. Hitherto IBM, because of its market dominance, has been able to impose its operating systems as the *de facto* world standard, while also fragmenting the market to insulate each area. The position was cumulative in the sense that as IBM hardware shares increased, so too did the number of applications designed to run on IBM operating systems, thus making it easier to sell IBM machines. IBM was then, in effect, able to lock customers into its own hardware – an important element in the high margin IBM was able to realize – and to control the rate and direction of product development (and hence the market itself). Rival firms had a choice: they either followed IBM or developed their own operating systems. The DEC

Table 6.3 European shares of computer operating systems, 1986 and 1991 (est.).

	1986	1991
Others	41	23
IBM 370	26	24
IBM S/3X	4	2
DOS*	18	23
DEC VMS	5	6
Unix	6	22

* including OS/2
Source: IDC market survey.

solution, to provide operating-system compatability across its product range, is in this sense a compromise between proprietary systems and open-systems. Many firms simply developed machines based around IBM (clones) to compete at the margins of IBM territory; Amstrad is a good example. Indeed IBM actually benefits from these firms in that they extend the dominance of IBM standards while IBM itself still has the corporate resources to maintain technological leadership. Others developed their own operating systems, usually designed with niche markets in mind. Firms such as Wang, Nixdorf and Apollo prospered in the early 1980s largely on the basis of their particular architectures.

Nonetheless, Table 6.3 does reveal the expected increase in the importance of Unix operating systems. Unix, developed by AT&T, is an attempt to create compatibility between the computers of different manufacturers: a global computer operating-system standard. The pressures for a unified operating system are immense; it is a fundamental condition for convergence to occur and it is being promoted by state contracts in the USA and Europe. However, the issue goes right to the heart of competitive relations in the industry, and unsurprisingly the creation of a global unified operating system is riven by corporate disputes.

On the one hand there is Unix International, a consortium led by AT&T and Sun Microsystems. On the other is the Open Software Foundation led by IBM and DEC. While the industry as a whole has been split between the two factions, both claim leadership in the development of the next generation of Unix. Clearly the major established computer hardware firms have an interest in maintaining proprietary systems in order to protect their installed base, particularly as most of their business comes from re-orders from established clients. They are also concerned about AT&T gaining the royalty rights from Unix, as an important source of revenue for the hardware firms lies in their operating systems. For the smaller hardware firms, open systems represent a double-edged competitive weapon. Open systems make it more possible to penetrate markets previously locked into competitors' technology, but also make their own installed base more vulnerable to competitive threats. Nonetheless the smaller niche firms must embrace Unix and open systems as a guarantee of customer security. This is because of the importance of upgrades and maintenance in computer systems. If a corporate user invests significant sums of money in a strategically important information system, it needs to feel assured that its purchase will be supported in the future. Because of their size and focus, niche firms cannot guarantee the long-term security that larger firms are able to. But open systems means that, should a niche firm fail, the user-firm is able to call on other suppliers without a complete system re-design, with all the expense and difficulty that entails.

The development of open-systems capabilities is by no means straightforward. Again, the niche firms with proprietary architectures have been the most vulnerable because they lack the scale required to overcome the resource costs involved in developing new open systems, while continuing to support previously successful proprietary systems. In a sense they are prisoners of their previous success.

Structural changes

As a consequence of these major changes in underlying technology, the applications of computers, and the demand for networking which underpins the demand for open systems, the 1980s have seen a dramatic restructuring of the world computer industry. The US computer industry lost some 40,000 jobs in 1988 (*Financial Times*, 12 December 1989, p.28) as a consequence of this process. Among the mainframe manufacturers, Unisys, Prime, Control Data and Data General have all recently announced losses, closures, redundancies and rationalizations. US firms have experienced problems in the super-computer market (*Financial Times* 18 April 1989, p.29; 27 April 1989, p.36) while others such as Honeywell and National

Semiconductor have effectively renounced their interest in computer manufacture. IBM itself has been struggling with flat sales and earnings growth, especially in the US market which, while still the worlds' largest, is growing much more slowly than that of Europe or Japan.

As noted above, a number of US and European firms that rocketed to prominence in the 1980s have subsequently collapsed with equal rapidity. Wang went into loss by the end of 1988, and by mid-1989 announced cuts of 2,000 jobs out of its 28,500 strong workforce, including the closure of its plant in Stirling, Scotland. Nixdorf similarly collapsed in 1988, by 1989 it had cut 1,699 jobs out of 31,000 and in early 1990 a controlling 51% interest was sold to the major German firm, Siemens. Zenith sold its entire PC business to Bull after a net loss of $15 million in the first quarter of 1989. The once high-flying Norsk Data of Norway has also plunged into loss and announced redundancies and restructuring. What Wang, Nixdorf and Norsk Data had in common were proprietary operating systems and, in the case of Nixdorf and Norsk, a geographically and sectorally restricted market. Most European computer hardware firms failed to appreciate the shift to PCs, with the partial exception of Olivetti, and are increasingly suffering the scale disadvantages of spatially limited markets – hence for example the purchase of Zenith by Bull and of Datachecker by ICL. A recent study also noted that European firms have been slow to develop open-systems capabilities (O'Brien 1989).

Other firms prospered in the 1980s, notably Sun Microsystems (workstations), Compaq (PCs), and Psion (laptops). Sun uses RISC components technology and open-system architecture, and has so dominated the workstation market that the two other large firms in the sector had to combine their interests, with Hewlett–Packard buying Apollo. Sun has been the fastest growing firm in US history, though there are signs that this phenomenal growth rate is now slowing down. Compaq too has grown fast; ironically, in the UK it has taken over the Stirling plant from Wang. The Japanese appear rather better positioned for laptops than they were for PCs: Toshiba already has a world lead and Fujitsu has bought a 38% stake in the US firm Poquet. Europe conspicuously lacks dynamic firms in these leading sectors of the computer hardware industry, and probably only Olivetti has the market presence to make a substantial impact in these product lines.

The major Japanese firms such as NEC, Fujitsu, Toshiba and Panasonic (Matsushita) have prospered in a relatively protected domestic market, but a combination of cost-push factors and the need to be close to markets is driving a more aggressive globalization strategy. Certainly the possibility of Japanese firms "hollowing-out" those in Europe is a danger; ICL is perhaps the most obvious case of manufacturing capacity being lost to Japan through its acquisition by Fujitsu, which allows the Japanese company a low-profile market-entry route. The competitive strengths of Japanese firms lie in components technology and manufacturing expertise, though they are also seeking to redress current weaknesses in computer architecture and software.

Markets are changing rapidly too. Two main features worthy of comment are the "global solutions" business and the mass market.

The "global solutions" business

One of the key shifts in consumption norms in the 1980s has been the desire of very large market firms and organizations to integrate their information technology strategies into a coherent whole which adequately addresses corporate strategic needs (see Morgan & Davies 1989). Computer hardware firms have been positioning themselves as "systems integrators", able to provide global solutions to corporate information technology requirements – though as the chapter on services makes clear this is a highly competitive market in which the hardware firms may not be the best placed to succeed.

Nonetheless the "global solutions" business requires computer hardware firms to be of sufficient size and geographical spread, and to understand clients needs in some detail. It also entails the partial customization of systems to client needs and a continued high level of support and maintenance. A further complication is that the user firms are seeking to bundle the servicing and maintenance of diverse computer systems into one package, handled by one overall supplier. Thus hardware firms need to develop the capabilities to maintain and upgrade other firms' equipment.

Firms of insufficient scale to become generalist system integrators are positioning themselves within particular vertical markets such as financial services systems or retail systems, a strategy for which ICL is a good illustration.

The mass market

As processing power becomes cheaper and applications more widespread, computer uses have penetrated deeper into advanced capitalist economies. In this sense computers have become low-margin commodity items where a premium is placed on high-volume, high-turnover sales. Globally this is an intensely competitive market (Table 6.4). While only about 15% of PCs are networked currently, that rate is expected to double over the next three years (*Financial Times,* 27 August 1989, p.11).

Table 6.4 Worldwide PC
shipments, 1988 (%)

IBM	11.5
Commodore	10.7
Apple	6.4
NEC	4.7
Amstrad	4.5
Zenith	3.0
Tandy	2.7
Compaq	2.7
Hewlett–Packard	0.6
Others	53.2

Source: IDC.

One consequence of the emergence of this mass market is that the hardware firms have to rely upon specialist and high-street retail outlets, rather than dealing direct with corporate data-processing departments. IBM has a very large number of agreements with "value added retailers", that is, retailers who add appropriate software and services to IBM products to suit them to particular markets.

Computer hardware firms in the UK and France

In the UK and France many people are employed in the computer hardware industry, as Table 6.5 illustrates. However, national policies towards the industry, and the character of that industry, are somewhat different. The sample of firms somewhat reflects these differences, which are of three main types: the national champion model and the free market model; the significance of US firms; and the growth of indigenous niche firms.

The national champion model and the free market model
France in the 1980s used state intervention and ownership overtly to protect indigenous firms and encourage the attainment of sufficient scale to compete in world markets, not only in computer hardware but many key sectors. Conversely, in the UK no attempt was made by the state to protect ICL in the 1980s. The firm's fortunes rested entirely on its ability to survive in a "free" market. This basic policy difference has helped create two very different types of indigenous hardware firms. On the one hand, Bull remains relatively dependent on the French market, but it has used an

Table 6.5 Computer hardware industry employment, 1981–84.

	1981	1984	1987
UK	56,799	71,225	70,230
France	44,817	51,520*	53,168

* 1983
Source: Department of Employment; NAP.

an aggressive acquisitions policy both to fill out its product range and to attempt territorial expansion. Its market positioning is far from clear (in terms of vertical markets) in part at least because the firm remains product-driven, functionally and hierarchically organized and bureaucratic in character.

The 1980s have been much more traumatic for ICL than Bull, but the firm may well be better placed as a result. ICL has pursued an aggressive rationalization and cost cutting policy. Overall the STC–ICL workforce fell by 31% in five years to 34,900 in 1987, with more cuts announced since, while vertical disintegration of production has accompanied the spatial concentration of assembly in the lower-cost North West.

The significance of US firms
Our sample of hardware firms includes three large US firms. This is a reflection of their importance, especially in the UK, in employment and production terms. The UK appears to be the preferred location for US and Japanese FDI in general, where free-market policies and an unwillingness to protect indigenous firms have underlined the abandonment of the national champion model. IBM (UK) is now the country's third largest equipment exporter, and employs 18,000 people directly. Employment in DEC (UK) has risen from 2,877 (1981) to 7,966 (1988) and contributes 28% of DEC Europe income. Hewlett–Packard employs around 18,000 in Europe of which 3,643 are in the UK and 3,586 in France. The US firms are already well established in Europe with integrated production and distribution networks. All three US firms have substantial R&D capabilities in the UK, and are well positioned for the European market after 1992.

The growth of indigenous niche firms
Within the UK there are some small computer hardware firms operating in particular market niches. Opus, for example, with a turnover of £30 million in 1989, specializes in IBM clones for office and word-processing

applications; Research Machine (1989 turnover of £50 million) sells PC-based CAD systems; Apricot (1988 turnover of £46 million) specializes in high performance workstations, though it now is dependent on Mitsubishi for hardware and specializes more in services–software and system integration work. Perhaps the most successful of the recent UK hardware firms has been Amstrad, which has built its reputation on very low cost IBM-compatible machines for the business and personal markets. Ironically Amstrad was in danger of breaching EEC rules on local content – its machines are produced largely in cheap labour locations in Asia – and has recently taken over the GPT plant at Kirkcaldy as its first UK product base.

None of these firms looks likely to equal the rapid growth of such US firms as Compaq. Moreover, the entry of Japanese consumer electronics firms into the European market (for example, Sony recently announced it would enter the workstation market) can only intensify competition still further. Amstrad held an estimated 10% of the European market in 1987, but with very short product life-cycles and low margins the transitions between success and failure can be rapid.

Three French niche firms were included in our sample. They are interesting for the rather different approaches taken towards niche markets. The example of Leonord is interesting for several reasons, since during the 1970s this SME has moved from the status of an electronics research laboratory to that of a specialist microcomputer enterprise. From the early 1980s the firm has grown very rapidly, thanks to public procurement in education and administration. However, the firm found it difficult in developing outside these markets and in 1986 it was acquired by IN2 and, in turn, both were acquired by Siemens in 1989. The case of Normerel is different. Having been the first French firm to produce IBM-compatible PCs in France in 1977, it was also first to develop more advanced compatibles (PS/2) in 1988, but technological sophistication and a distribution agreement with the Taiwanese firm Arche Technology were insufficient to wipe out past losses. As a consequence in 1990 Normerel was acquired by SMT Goupil, the second-ranking French firm.

SMT was established in 1979 as a microcomputer specialist. However, the firm does not have its own production facilities. The company has pursued a policy of acquisitions, for example purchasing Sféna, a division of Aérospatiale. In 1987 SMT was the subject of a management buy-out to avoid a public takeover. The purchase of Normerel gives SMT Goupil a production base in Normandy and it facilitates expansion into the international market. Nevertheless, the strategy of growth by acquisition has not worked. In 1991 SMT Goupil filed a loss of £45 million and is itself seeking a buyer (possibly Olivetti) for its 750-employee operation.

Internal organizational dynamics

Here developments in the internal organization of our sample firms are considered, chiefly with respect to a) the shift from functional to vertical market organization focus, b) micro-regulatory changes in labour relations, and c) spatial configuration. These changes are driven by firms seeking to offer "complex products" which require closer organizational integration of functional operations from R&D to marketing. Table 6.6 summarizes our sample firms in terms of internal and external organizational dynamics.

The large, broad-capability firms are determined to remain volume producers of computer hardware. IBM clearly has an impressive lead already simply by virtue of its size. Nonetheless all these firms have implemented major internal reorganization in the 1980s. Typically, the IBM reorganization has been the most striking. Following a profits collapse in 1986 (in part caused by the costs of trying to acquire and then manage communications capabilities) the company adopted a three-year austerity and refocusing programme in which 16,000 jobs were lost (from a total of about 400,000) and 21,000 transferred to "front-line" (i.e. marketing) jobs. Further cuts have followed: in the USA in 1988 10,000 job losses were announced, with the closure of several manufacturing plants. In parallel with this process, the corporate structure has been transformed into seven major business lines, ranging from "applications solutions" (i.e. systems integration) to "personal systems" (i.e. PS/2 and RT workstation, plus appropriate software). Moreover, the company has adopted the so-called EXCEL quality programme across its operations, with significant implications for the organization of work. Such changes have obviously been felt in the UK and France too, with some reduction in employment levels and a far greater emphasis on monitoring performance at a disaggregated level (see, for example, Doran 1986). IBM has a truly globalized production system, such that certain locations may be the world (or continental) source for particular products. From the UK, for example, Greenock (Strathcyde) is a continental source for PS/2 machines, Havant (Hampshire) a world source for the 9335 disk files, and Hursley (Hampshire) a world source for PS/2 colour monitors.

Hewlett–Packard ("H-P") similarly adopted a major reorganization programme (under consultants McKinsey) in 1984 which transformed scores of divisions into five market-based key business sectors markets, of

Table 6.6 Markets, specialization and strategy of sample computing firms in Britain and France. ("Full" in **bold**, "hybrid" in *italics*, "niche" in roman.)

	Specialization in computers	Specialization within computers	Vertical integration	Service diversification	Exports	Globalization	Acquisitions	Alliances	Organization	Strategy	Evaluation
IBM	Strong	Full, inc. peripherals & software	Strong esp. components	Yes, VANS	Non-US is 57% turnover. Has global production	High degree	Expensive failures in telecommunications	Vital in software. Many retail alliances	Shift from product to matrix; market facing	To be global solutions provider	Traumatic 1980s; partial recovery
NEC	Strong	Full, badge PCs	Strong but growing subcontractors	Yes, networks	Similar to IBM, less strong in Japan	High degree	No major buys	Vital in PCs (Tandy, Olivetti)	Flexible matrix; informal	To expand networks, esp. high quality & high cost	Successful
Bull	Strong	Full inc. NEC & Zenith	Strong	No	75% of sales are in Europe esp. France	Medium: expansion in UK and US	Key part of strategy e.g. Zenith	Vital with NEC for mainframes	Complex traditional structure	To become second IBM	Problematic; major changes needed
Fujitsu	Medium computers 66% of sales	Full range	Strong	Yes	Indirect e.g. through ICL	Medium: less direct investment	Increasing importance e.g. Poquet (US)	Vital for geog. expansions e.g. ICL, Siemens	Traditional Japanese	Manufacturing excellence and alliances	Successful
Hewlett Packard	Medium-low: information systems 54% of sales	Full-range esp. PCs & workstations	Partial	No	Has global production	High degree	Purchase of Apollo vital	No core alliances	Work teams	To become market facing, esp. workstations & networks	Successful but still high cost; lack of focus
ICL	Strong	Partial	Weak	Yes, esp. retail	Retail most export orientated; otherwise weak	Medium: expansion in US	Important in US e.g. Datachecker	Significant in components; also Fujitsu	Reorganization to flexibility resisted by unions	Reduce costs; target vertical markets	Successful but vulnerable to acquisitions
Nokia Data	Low: 23% of conglomerate sales	PCs, workstations, networks	Weak	Partial	64% sales in Scandinavia	Only in Europe	V. important e.g. Ericsson Info. Systems	OEM agreements to create volume	Rationalization of acquisitions	Develop electronics; expand to US	Inadequate integrations
Nonord	Strong	Micro & mini	Weak	No	Weak	Very low	Bought by IN2 and Siemens	None of significance	Product based	Absorbed by Siemens	Absorbed
Normerel	Strong	Micro only	Strong	No	OEM strategy: exports more important	Very low	Bought by Goupil	OEM agreements	Traditional functional	Target niche markets e.g. state education	Successful but potential now absorbed
Goupil	Strong	Micro & mini	Very weak only design & distribution	Partial	Weak	Very low	Bought Leonord	None of significance	Modular	Concentrate on high performance certain systems	Failure

which the three most important are "Measurement Systems", "Business Systems" and "Technical Systems". Traditionally Hewlett–Packard had been very much an engineering-led company with a relaxed management style, and a very complex and disconnected structure. The result was creativity, a vast array of products but also, increasingly, a lack of competitiveness and low productivity. More importantly perhaps, the ubiquitous nature of computers made it increasingly important for H-P to understand consumer needs more closely. Thus, like IBM, H-P has tried to create an organizational structure which increases market sensitivity. Hewlett–Packard is also a good illustration of the basic organizational difficulties facing computer hardware firms. On the one hand they need to encourage creativity and flexibility among their staff, and often have very distinctive corporate cultures which reflect this basic condition (for example the "H-P Way") (Knobel 1988, Maxon 1988). On the other hand a strict control over quality is necessary. In the factory the H-P solution has been "work teams" (approximately 15 people) which meet weekly to discuss performance, together with the more rigorous adoption of performance-related pay. Management sets performance standards and goals, but a substantial autonomy remains in terms of how those goals are to be met.

DEC has long been a strong practitioner of "open-door" management, and also has a well developed matrix management approach which is well supported by DEC's own networking skills. DEC is highly integrated both in product terms (from components to sub-assemblies to computers) but also in functional terms (R&D, production, sales, etc.). The company successfully transformed itself from a product-led niche firm to a broad capability computer hardware firm (Thackery 1987), before the "PC squeeze" in 1989–90.

The position for Bull is far less hopeful. As noted above, Bull is the product of the "national champion" policy in France: the French have been concerned to have an indigenous computer industry since the USA blocked the sale of computers needed for atomic weapons research in the 1950s. The entire Groupe Bull revenue of $5 billion makes it the world's ninth largest computer hardware firm, but this may be insufficient to guarantee long-term viability. The 1990 turnover of Bull, including Zenith, was £350 million. Estimated losses amounted to £68 million with the French state announcing a recapitalization of £20 million. Bull is involved in massive restructuring and workforce rationalization. Between 1990 and 1991, 8,500 jobs were eliminated, a cut of approximately 20% in the workforce, and the number of production sites was reduced from thirteen to seven. Bull is still organized along functional lines, although its aggressive acquisitions policy has facilitated global expansion and the filling-out of its product line. Nonetheless the firm retains the appearance of a smaller and somewhat out-of-date IBM. It has production difficulties within France, especially in Angers where union influence is strong, and has recently been forced into (somewhat belated and reactive) rationalization with 1,200 job cuts announced in 1990 following on from 1,000 "natural wastage" reductions the previous year. In all 6.5% of the French workforce has been cut. In other words Groupe Bull appears out of step with the rest of the computer hardware industry, a legacy of overt state intervention and protectionism. The price of global expansion must be domestic retrenchment, but the consequent political implications will make the transformation difficult.

The hybrid firms ICL and Nokia have adopted very different approaches to the computer hardware industry. ICL, created out of the 1960s Wilson era of industrial concentration (Kelly 1987), suffered a catastrophic loss of market share in the 1960s from which it has never fully recovered. Its multi-plant, multi-product and multi-locational inheritance was unwieldy, with significant duplication and overlap. Since its inception then, ICL has been concerned to rationalize this heritage, and has abandoned any pretensions towards being a large-scale, higher-volume, computer hardware firm. The rationalization process has accelerated in the 1980s, in parallel with a sustained attack on trades union barriers to flexibility. Indeed, for a while in the mid-1980s, ICL was seen by many as epitomizing the changes which UK manufacturing industry in general had to embrace if it was to survive into the 1990s (Caulkin 1987, Clarke 1986). However, the process of achieving labour flexibility has not been straightforward, especially in the context of continued redundancies, and it may have cost the firm in terms of defections of key staff to other firms. ICL now has only three manufacturing locations: Ashton near Manchester, Kidsgrove (making PCBs) and Stevenage (software development, warehousing and distribution). Ashton, using ICL technology, is regarded as a "showpiece of flexible manufacturing", but it is more of an assembly and logistics operation. ICL has become a highly disintegrated firm. It was also one of the first firms to realize that globalization plus a vertical market niche strategy was a possible survival route in the face of competition from such industry giants as IBM, DEC, etc. Alliances and acquisitions have been important here, as the following section on external organizational dynamics shows.

Nokia has a very different history and approach to computer hardware. Nokia has long been the largest company in Finland, a broadly based conglomerate with its corporate roots in basic industries such as wood, paper and mining. A corporate diversification and geographical expansion plan was put into place in the late 1970s in which electronics had a key rôle: it rose from less than 10% of total Nokia income in 1980 to over 60% in 1988 (see Vliet 1987). In the process Nokia has developed a substantial presence in mobile telecommunications, in consumer electronics (purchase of SEL) and in information systems (purchase of Ericssons' data communication interests). Information systems contributed 23% of net sales in 1988. The firm has become European rather than Finnish (87% sales in EEC/EFTA) but apart from mobile telephones has yet to become global. However, it has successfully used its financial and technical resources, plus its strong management team, to focus attention on relatively specialist niche markets which it is now seeking to expand, for example in EFTPOS systems, banking terminals, etc. The expansion into electronics has not been without problems. The acquisitions strategy has presented problems of integration and management, and there still remains a question mark over the relative lack of scale of Nokia. Its strategy in the data communications division has been to concentrate on open-systems high-performance workstations in partnership with specialist software houses such as 3Com. Nokia Data and ICL merged in 1991.

External organizational dynamics

External organizational dynamics refers to the relationships between a firm and its constituent world (largely, of course, other firms). Clearly, there is a close link between internal and external organizational dynamics; for example, a shift towards vertical disintegration and assembly, as noted above, also implies increased subcontracting. Similarly, a shift towards "core competences" implies the need for other (external) means of gaining access to the required competences for systems integration, networking and so forth, either by purchase on the open market or, and more probably, by various forms of alliance. This final section therefore considers the external organizational dynamics of the computer hardware industry in the 1980s, again based on our sample firms, paying particular attention to alliances.

It is clear that alliances take many forms, only some of which can properly be described as quasi-vertical integration strategies. There are not always clear distinctions between alliances and other forms of corporate strategy, notably in terms of acquisition. Often alliances in the computer hardware industry are indeed cross-sectoral, what we may term competence alliances; but others are internal to the sector (ICL–Fujitsu) and are concerned with economies of scale and geographical market penetration as well as technological synergy.

In general, the hardware industry is externalizing much more of its routine and subassembly work, chiefly in an attempt to keep down costs in an increasingly low-margin product. But the rise of systems integration work is also driving the trend towards higher quality on the hardware side, and forcing the hardware firms to devote more resources to higher value-added activities such as information-systems consultancy, customization and so forth.

Of the major hardware firms studied, DEC is probably the most integrated of all. IBM is of course one of the world's leading producers of electronic components, and it derives a substantial proportion of its income from software (13% in 1988); nonetheless the firm has been increasing the volume of subcontract work while reducing the number of subcontract suppliers. In the UK the number of suppliers fell by over 28% from 6,720 to 4,796 in just one year between 1986 and 1987 even though expenditure on subcontract suppliers jumped from £720 million (1986) to £920 million (1987).

The number of IBM alliances has been proliferating; including dealership alliances, they literally number thousands. As such it is impossible to give the full impression of IBM relationships with other organizations; however, a number of genuinely "strategic" alliances are evident. First, there are the links with the software houses vital to customize IBM machinery to particular applications. Firms such as Microsoft (responsible for MS–DOS and OS/2), Management Science America (IBM has a 5% share) and Comshare are of great importance at either a global or regional level. Secondly, there are the telecommunications linkages which appear to have crystallized around Siemens, but which extend to network services operations such as the Paribas VAN in France. Thirdly, there are the standards agreements, most notably the Open Software Foundation, which rest somewhat uneasily on the IBM legacy of market and product fragmentation. Finally, there are a large number of market-specific alliances in which IBM is simply using other organizations to achieve greater penetration; the dealership alliances fall into this category. Another example could be the alliance with Robertshaw construction to create an "intelligent buildings" complex, much of which is to be used by IBM.

Apart from standards agreements, IBM has tended to remain somewhat distant from the rest of the computer hardware industry (though it has recently announced a cross-licence agreement with Compaq). Its acquisition and alliance activities have concentrated on telecommunications and software, with smaller piecemeal arrangements with a wide variety of firms for particular applications. This model seems typical of the broad capability hardware firms in our sample except of course Groupe Bull, which as Honeywell Bull was conceived of as an alliance between three computer hardware firms (Honeywell, Bull and NEC). The fate of the Honeywell Bull alliance is perhaps an interesting illustration of the dangers of alliances within an industrial sector. Bull regarded the alliance as a stepping stone to internationalization via the Honeywell customer base; Honeywell has now been pushed aside almost completely. The NEC part of the alliance (which provides the top-end processors) has been kept rather quiet, and it must be feared that they, like Fujitsu, are undertaking a hollowing-out strategy: in other words, alliances within a sector tend to lead to takeovers and mergers with the eventual dissolution of one or more parties as a separate corporate entity.

In both their alliance and subcontract activities, the major hardware firms face difficult choices as to which elements constitute the core technologies that must remain under direct control, and which can be "safely" entrusted to suppliers or collaborators. Increased vertical disintegration makes the final assembler–producer potentially more vulnerable to disruptions in the supply of products and to increased dependency upon others. The greater complexity of external organizational linkages in both product and market terms leads to managerial and scheduling difficulties of potentially crucial importance in the rapidly changing market conditions faced by the major hardware firms.

Hewlett–Packard's most significant move has been the purchase of Apollo, a rival in the workstation market. In 1988 the global market for workstations (estimated at $4 billion) was shared as follows (*Source:* IDC):

	%
Sun Microsystems	28.3
DEC	18.6
Hewlett–Packard	16.9
Apollo	13.5
Silicon Graphics	4.4
Others	18.3

Apollo was the first major Hewlett–Packard purchase (at $500 million), having itself fallen into loss in 1987 as a result of its failure to embrace open standards. It is an example of the importance of more traditional corporate strategies alongside dynamic alliance activity. Most of H–P's alliance and minority ownership activity is concentrated on software and network firms (e.g. 10% of 3Com, 10% of Octel, 40% of Estel, alliance with Northern Telecom for networks), but it does not appear to have developed the linkages to markets and users which IBM has.

DEC, being somewhat smaller, has been more open to other computer hardware firms. DEC has been noticably weak in developing smaller machines of its own to compete in the PC market, and has therefore introduced agreements in which other specialist manufacturers have adapted their machines for DEC networks: for example, Apple (1987), Olivetti (1988), Tandem (1987), Tandy (1988) and Compaq (1988). DEC networking skills are very well developed compared with much of the computer hardware industry, and this type of alliance produces a reasonable fit between the corporate aspirations of the parties without threatening the heart of DEC's business. Indeed DEC developed the pioneering "Ethernet" network standard in alliance with Xerox and Intel some years ago, and has link-ups with firms such as Stratacom, Timeplex and Motorola Codex to develop further network applications. Thus, alliances here are used to fill out the product line within specific geographical markets. As with other large computer hardware firms, DEC has been developing greater levels of subcontracting (and an emerging EDI network) in low-value components (printed circuit boards, printers, other peripherals) and in "badge" products from such firms as Olivetti and Tandy. DEC is also interesting in that it has tried to apply its own CIM techniques used in large-batch production of computers to other areas, notably in association with Comau (Fiat) and Allen–Bradley.

Both of the hybrid firms in our sample have undergone profound external organizational restructuring in the 1980s, though in very different ways. On the one hand, Nokia have pursued an aggressive acquisitions policy and have boosted manufacturing volume through OEM agreements with larger firms (e.g. Nokia supplies terminal screens to IBM). Total employment in the firm has grown from 28,500 in 1986 to 46,000 in 1988; at the same time the firm has developed a European structure with divisional headquarters outside Finland (the headquarters of Nokia Data are in Stockholm). The strategy of acquisition and integration is matched by a pragmatic approach to alliances and offshore production. In telecommunications, for example, Nokia has a world lead in handset sales, chiefly due to its marketing alliance with Tandy in the USA, with low-cost components supplied by a joint venture in South Korea. In Europe, where political complexities and the next generation of digital mobile telecommunications complicate the picture, Nokia has formed an alliance with Alcatel and AEG. ICL by contrast has pursued an aggressive rationalization policy, cutting back on actual manufacturing and developing a more European supply base. Acquisitions have been used to extend vertical market presence on a global scale, most notably the purchase of Datachecker from National Semiconductor, but it must be admitted that ICL (and indeed STC, its former parent) has always been more the subject of takeover speculation than the instigator. Its relative dependence on Fujitsu, for example, explains why in 1990 the Japanese firm purchased an 80% equity stake in ICL. Despite fierce cost-cutting, ICL had not produced results strong enough to quieten takeover speculation, in part at least because of increased pressure in its strongest domestic markets in local government, health and education.

Conclusions

Computer hardware firms of all types have been challenged by the technological and marketing difficulties of convergence in its crude form and in its more recent and complex "systems integration" form. In general, after some expensive mistakes, the response has been to concentrate upon core competences, reduce in-house production, drive up quality via preferred suppliers and in-house quality management programmes, and develop more flexible external organizational relationships with firms able to provide complementary capabilities. These other firms may be in components, other computer hardware, software, network managers, consultants or major users (and retailers). For the niche and hybrid firms, globalization and vertical market focus appear most likely to offer an independent future, while the arrival of open systems offers both threats and opportunities. It may prove impossible for the "quality" hardware firms to exist in the mass market, but equally, as Chapter 7 shows, many firms from a range of sectors and organizational traditions are positioning themselves to capture the emerging systems-integration market.

CHAPTER SEVEN
The telecommunications equipment industry: the great transformation

Olivier Weinstein

Following more than fifty years of evolutionary development under stable structural conditions, the telecommunications industry has undergone rapid structural changes since the middle of the 1970s. This transformation has been driven primarily by technology. The transition from electromechanical to microelectronic technology and digitalization has led to a radical change in production processes and a multiplication of new products and services, transforming the telecommunications industry from a single-product sector into a complex diversified sector, converging with the computing industry. The change has affected all types of equipment, terminals, switching and transmission systems.

Technological change has stimulated the expansion of new market potential and the transformation of institutional and market structures. Until the end of the 1970s, in France, the UK and other European countries the telecommunications industry was highly regulated. It was based on a public monopoly in services supply, and a system of special relations between the PTT administration and a small number of equipment suppliers. The telecommunications equipment firms were thus producing mainly for a national public demand. The diversification of products and services and the internationalization of telecommunications markets have made the existing institutional arrangements increasingly inadequate. Adaptation to the new context has differed from country to country. In particular, France and the UK have moved in almost opposite directions: a system of "mixed economy", with a reorganization of the public monopoly and an active industrial policy in France, and liberalization and privatization in the UK. The consequences for equipment suppliers have been critical.

We will see in this chapter how technical, institutional and market transformations have challenged the position of telecommunications firms, and how firms are seeking to meet the new challenges.

Technological and institutional changes

Process and product innovations: the new technical system

The new telecommunications technical system is the result of converging innovations (Ungerer & Costello 1988): developments in microelectronic components; digitalization of switching and transmission, which reduces costs and allows the unification of voice, data and image transmission and processing; the substitution of electromechanical control systems by computer systems – stored program control (SPC) – which makes software the central part of the switching system, offering greater versatility and opportunities for new "intelligent" functions; new transmission systems – optical fibres, satellites and microwave technology – providing greater transmission speed and capacity, and permitting the development of new services and of low-cost long-distance links. Telecommunications systems are becoming computer systems: the digital exchanges are large specialized mainframes. This development provides the technological basis for convergence of computers and telecommunications technology, generating new services and products, and the possibility of "intelligent" networks and terminals, leading to "integrated services digital networks" (ISDN).

The impact of new technology on process and products is considerable. The production process is transformed in different directions (Table 7.1). There is a prominent rôle for electronics components and, above all, for software, which is the critical part of new switching systems, leading to a restructuring of production processes and of the knowledge-base of the industry. A new employment and skills structure is evident, with a rapid decline of direct labour and a growing proportion of engineers and technicians. And finally, there is a shortening of equipment lifetime, with a very rapid increase of R&D costs. On the whole, the change can be analysed as a transition from a Fordist production system to a knowledge-intensive or science-based one.

The most important change, however, lies in the multiplication and growing sophistication of services and products. The service dimension will

Table 7.1 Transformation of the production process for switching systems

(a) Direct labour in switching manufacturing (50,000 lines p.a.)

	Crossbar	Semi-electronic	System X
Material & metal-working	1,000	150	20
Assembly & cabling	2,000	900	50
Test	250	200	50
TOTAL	3,250	200	50

Source: Corfield (1978), in de Olivera (1981).

(b) Cost structure of public switching systems (estimated)

	1970	1990
Software	20%	80%
Hardware	80%	20%

Source: Ungerer & Costello (1988).

(c) Characteristics of public switching systems

	Digital system (end of 1980s)	Electromechanical system (early 1970s)
R&D costs	1 billion ECU	15–20 million ECU
Lifetime	10–15 years	20–30 years
	Continuous product updating, cost 100 million ECU p.a.	No important modifications

Source: Ungerer & Costello (1988), Roobeck (1988).

(d) R&D costs of digital switching systems

Company	System	Estimated costs (US$ million)
ITT	System 12	1,000
Ericsson	AXE	500
CIT–Alcatel	E10 & E12	1,000
Northern Telecom	DMS	700
GEC/Plessey/BT	System X	1,400
Western Electric	ESS-5	750
Siemens	EWS-D	700

Source: Roobeck (1988).

be analysed in the next chapter. On the equipment side, until the 1970s the range of telecommunications products consisted mainly of a few large public-equipment goods (switching and transmission systems) and simple terminals (telephone and telex). The new technology generated an increasing range of terminals (for voice, text, data and image) and of new systems, such as cellular telephony, private networks, or new generations of PABX. This evolution has two main directions (Ungerer & Costello 1988): integration of transmission and processing functions, and increasing scope for user equipment, an important outcome of digital technology being that an increasing number of functions can be realized outside the network itself, on equipment connected to it (Roobeck 1988). The result has been a complete change of market configurations. But let us look first at changes in the institutional framework of the telecom sector.

Institutional changes: mixed economy versus neoliberalism

Changes in the institutional framework are as important as those in technology to explain the evolution of markets and firms. Until the 1970s institutional environments for telecommunications in the UK and France were similar. As in other European countries, the telecommunications equipment industry was characterized by close relationships between the public network operator and a "club" of equipment suppliers. In both countries, public administration played a prominent part in technology and R&D policy, in particular in the development of digital switching systems. But, from the mid-1970s, institutional regimes in France and the UK have been moving in opposite directions.

France

As yet, there has been no real deregulation in French telecommunications. France Télécom (the ex-DGT) remains a department of the French Post Office. However, that apparent continuity conceals an important change. At the beginning of the 1970s the French telecommunications infrastructure was very backward, with one of the lowest telephone densities in Europe. The supply of equipment was dominated by foreign firms. The 7th Plan stimulated an ambitious telecommunications policy, combining a restructuring of the equipment industry with a transformation of the DGT (cf. Barreau & Mouline 1987). There were three targets: fast expansion of the telephone network, the creation of a strong French telecommunications industry, and the growth of exports. The DGT became the first French investor; telecommunications investment remained at a high level from 1975 onwards, and far above that of British Telecom (Table 7.2).

Table 7.2 Investment in telecommunications operations: 3-year moving average (constant 1980, US$m).

	1975-77	1977-79	1979-81	1981-83	1983-85
France	4521.95	5598.16	5190.54	4566.97	4535.04
UK	2967.63	2494.53	2746.40	2837.55	2792.07

Source: OECD (1988).

The reorganization of DGT has transformed it into a commercially orientated business firm, even if it remains formally part of public administration. A more business-like management system has been set up, the structure has been regionally decentralized, and the DGT is financially autonomous. The relationships between the public operator and the equipment suppliers have also been modified. These modifications were made with different goals in mind: competitive tendering to develop competition, and promotion of industrial R&D by financial and technical support of the CNET, the DGT research department, which has kept the leading rôle in the R&D policy.

By the end of the 1970s, the telecommunications scene was totally transformed. The domestication of the equipment industry was almost complete, with CIT Alcatel and Thomson covering more than 70% of the public switching market. CNET and Alcatel were early to perfect a digital exchange. The diffusion of electronic and digital technology proceeded rapidly, France having today a highly advanced digital telecommunications network.

During the 1980s, using its lead in digital technology, the DGT has expanded its supply of networks and new services by the creation of subsidiaries. With regard to terminals, the DGT policy has always been relatively liberal, more than in the UK before privatization (Morgan 1989). Today, all the markets for terminals are open to private suppliers, as is the case in the UK. But France Télécom still retains control over market entry by means of an agreement procedure. The orientation with respect to value-added network services remains uncertain, but liberalization has begun with, for example, two networks for cellular mobile telephones, one managed by France Télécom, and a private one managed by the SFR (Société Française de Radiotéléphone). The new law on telecommunications regulation (October 1990) controls competition under three "regimes": public monopoly for the telephone networks; competition subject to agreement for radio-communications, private networks, data transmission and terminals; and free competition for other value-added services.

The main telecommunications firms, Alcatel with CGE, Thomson and CGCT (an ITT subsidiary), were among the firms nationalized in 1982. This had one paradoxical effect: to make them more powerful and autonomous (Morgan 1989). The new trend of socialist policy after 1983 and the denationalizations of 1987, in particular of the CGE group, have reinforced that autonomy. The restructuring in the second half of the 1980s was managed mainly by the firms themselves, and led to the supremacy of Alcatel, as will be shown later. Thus, during the internationalization process of the 1980s, one of the most important developments has been the growing independence of firms (public or private) from public administrations. This evolution is consistent with a traditional goal of French industrial policy: to create "international champions".

Overall, the dynamism of France Télécom and the expansion of Alcatel can be seen as the two sides of a same system, a system of "mixed economy".

The UK

The evolution in the UK has been radically different in many respects. A first difference with France, before 1983, concerned the policy of British Telecom and its relationships with equipment firms. In the 1970s the telecommunications business in the UK was widely regarded as being not very efficient: it was technologically backward and BT's productivity was low. The problems arose from poor management of the Post Office and British Telecom, but also from the failure of the equipment industry (Morgan 1989). The government and the Post Office failed to transform the system as in France in the same period. The difficulties and cost of the development of System X illustrated all too well the inappropriate relationship between the Post Office and the firms of the "telecom club". The political influence of the main firms seems to have prevented the rationalization of the industry. This was in marked contrast with France, where the relative weakness of company power, the tradition of *dirigisme*, and an active industrial policy aiming to create "national champions", have promoted a complete restructuring of the telecommunications industry.

Until the 1970s the UK telecommunications market was probably the least open in Europe. Between 1979 and 1984 the institutional system was completely transformed, making the UK telecommunications regime the most liberal in Europe, combining privatization and liberalization of markets, by different measures, including:

- the creation, by a consortium of Cable & Wireless, Barclays Bank and BP, of a second network operator, Mercury, becoming a rival to BT in national and international markets;
- the privatization of BT;
- a partial liberalization of basic network services, with the creation of Mercury and the licensing of two consortia in cellular mobile telephone networks (led by BT and Racal);
- an almost total liberalization of terminal equipment and value-added network services.

The privatization of BT as an entirety was the most important measure. For the economists and the equipment manufacturers who were the main opponents of privatization, it was seen as a substitution of a private monopoly for a public monopoly (Encaoua & Koebel 1987, Morgan 1989). Actually, for the supply of basic services, the market structure is one of closed duopoly – not a very liberal situation at all. In spite of the entry of Mercury, the monopoly power of BT has remained very strong, probably becoming even stronger than before the reforms.

This evolution had a profound impact on the position of the telecommunications equipment firms. It radically changed the nature of the relationships between them and the public network operator: BT became a potential competitor. The politics of the new privatized firm has reinforced this impact, first by the decision to diversify its demand for supplies in the direction of international IT companies (Ericsson for public switching), then by its strategy of diversification in the telecommunications business, in particular in equipment through the acquisition in 1986 of Mitel, a Canadian supplier of PABX. In both cases, the transformation of BT and the new market configurations seemed to have weakened the position of UK equipment suppliers. However, they, in turn, have responded with price-cutting competition, which has resulted in BT seeking to sell its PABX acquisition, Mitel.

Overall, the British liberal policy has had contradictory effects. On one side, it has made BT more dynamic and has stimulated the growth of services, mainly to the benefit of large business users (Morgan 1989),but it has also had positive effects on the development of new (value-added) services, for example in the mobile telecommunications business. On the other side, it has put the UK telecommunications equipment firms in a very vulnerable position. In contrast to the French position, the UK trade balance for telecommunications is worsening (Table 7.3). But the weakness of the UK equipment industry cannot be explained by liberal policy alone. It has also come from the deficiencies of the firms, particularly in R&D, and the incapacity to restructure the industry in due time (the creation of GPT (GEC–Plessey Telecommunications) in 1987 coming probably too late).

Table 7.3 Commercial trade balance: telecommunications equipment, France and the UK (thousands of US$).

	1978	1980	1982	1984
France	301	371	433	445
UK	243	244	162	100

Source: OECD (1988).

The contrast in development of the telecommunications sectors of France and the UK is particularly pronounced. The telecommunications industry is one of the main successes of French industrial policy (together with aerospace), combining a strong and efficient public network operator and the formation of a powerful global firm, Alcatel NV. In the telecommunications equipment industry, the UK situation is far weaker. However, the process of internationalization, the pressures for liberalization of markets in Europe and the relative decline of public demand are changing the rules of the game. In that new context, the French institutional arrangement cannot remain the same: development will become more dependent on the strategies of firms, and, therefore, will be less dependent on the state.

Products and markets

Until the 1970s, telecommunications equipment firms produced and supplied mainly public systems (switching and transmission) and simple terminals (telephones and telexes) for public organizations, within the national framework. The markets were national monopolies. Technical and institutional changes have led to a radical transformation of telecommunications markets and their structures during the 1980s. The growth of equipment markets has been rapid (Table 7.4), but the most important evolutionary trends lie in structural shifts in three directions: the development of new products with a movement towards integrated systems, the "privatization" of markets, and the internationalization and globalization of the industry.

159

Diversification of products and development of integrated systems
Telecommunications equipment is usually divided into five categories:

–   public switching equipment, which remains the principal segment
      of the telecommunications equipment markets (Table 7.5);
–   transmission equipment, with the emergence of two new
      technologies, namely, fibre-optic cabling and satellites;
–   private switching equipment, with the important PABX markets;
–   terminals: telephone sets, telexes, videotext and teletext terminals,
      fax terminals, and complex new products such as multipurpose
      terminals and ISDN terminals;
–   cellular mobile communications equipment, which forms a new and
      fast-growing segment of the telecommunications market.

Table 7.4 Growth of telecommunications equipment markets (estimations) (average growth rate 1983-90, %)

	All equipment	Public switching	Transmission	Private switching	Telex	Data transmission
Europe	6.8	4.3	9.1	10.9	10.7	13.9
North America	8.9	8.4	13.3	7.4	7.8	16.6
Asia	7.4	4.3	10.1	10.2	7.7	20.7
WORLD	7.9	6.1	10.9	9.6	9.0	15.7

Source: Italtel, in OECD (1988).

Table 7.5 Telecommunications equipment markets in France and UK (US$ million, estimated)

	1987		1988		1989	
	France	UK	France	UK	France	UK
Customer-premise equipment	599	408	601	414	603	421
Telephone & data-switching systems, private	503	389	522	372	541	355
Telephone & data-switching systems, public	1,129	1,357	1,163	1,443	1,197	1,534
Transmission & carrier equipment	613	645	607	659	601	673
TOTAL	2,844	2,799	2,893	2,888	2,942	2,983

Source: Electronics, January 1989.

The evolution of the telecommunications equipment industry is first characterized by a multiplication of new products, particularly in terminals and business systems, with two main tendencies (Table 7.6). The first is that the public switching market remains the most important market, especially in the most industrialized countries, although its relative importance is declining. The highest growth rates are in new strategic markets: in business systems including new terminals (telefax, data systems), new generations of PABX, private wide-area and local-area networks, and mobile communications systems as well as joint products (relay, specific terminals, etc.) which constitute one of the most promising technological and commercial developments for the next decade. The second tendency is that, in terminals and private systems, the share of stand-alone products will decline with the rise of integrated business information systems. It means that markets are becoming markets for systems or "solutions", more than markets for specific products, and that the complementarities between products and services are expanding. The integration of products and markets is an expression of a general move to a new "industrial paradigm" (cf. Chs 1 & 2), connected to technical change (digitalization), the exploitation of economies of scale and scope in R&D and marketing activities, as well as the evolution of user needs.

Table 7.6 Growth of world telecoms markets towards integrated systems (%)

1980 90 billion ECU		1987 200 billion ECU		After 1993 300–400 billion ECU	
Public switching	11.5	Public switching and transmission	23	Public networks	20
Transmission	17.6				
Private switching	9.8	Private telephone networks (with data & text capabilities)	18	Integrated telematic systems	55
Telephone terminals	6.6	Private data & text systems	19		
Decentralized data systems	13.8				
Non-voice terminals	9.7				
Typewriters	5.1	Stand-alone products	42	Stand-alone products	25
Output systems/devices	11.4				
Storage systems/devices	14.6				

Source: Arthur D. Little in EEC (1987)

Privatization of markets

One of the most important outcomes of these developments for telecommunications equipment firms is the transformation of market mechanisms: a growing importance of private competitive markets relative to traditional public markets. In France and the UK the telecommunications equipment firms were orientated towards public demand, very often diversified in other similar activities such as defence electronics, aerospace or transportation equipment. This involved a specific business organization and culture, based on special long-lasting relationships with government agencies and the protection of strict institutional barriers to entry. It meant also that markets were extremely stable and predictable.

The business systems markets are totally different. Competition is strong, despite the fact that some markets remain protected by national norms (e.g. the French PABX market). The life cycle of products is becoming shorter, and the capacity to commercialize a large range of products fitting rapidly changing user needs has become crucial. The telecommunications equipment firms must also compete with the network operators supplying virtual networks, and with computer firms entering the markets by offering local networks.

Public or state-regulated markets are changing too. In the UK the privatization of BT transformed the market from public to private. For mobile communications systems in most European countries, the markets have been partly deregulated, albeit in different degrees. Competition, organized in international consortia, has stiffened. For the equipment firms, this new context leads to new organizational forms, more commercially orientated and more flexible, with new types of relationships with clients (public and private) and rival firms. On the whole, it means a total change of business culture, structure and behaviour.

In spite of the privatization and liberalization of telecommunications equipment markets, the rôle of public administration remains important. In most countries, public demand is still large, and public administration continues to control private markets by setting technical norms and agreement procedures. The difference between the UK and France (and most of the European countries) is very important in that respect. As we saw before, liberalization and privatization have weakened UK equipment firms. In France, even if all the terminal and private equipment markets are open to private suppliers, France Télécom imposes strict control on entry, and the firms can combine the advantages of steady public demand with the expansion of private and international markets. It is also worth noting that the politics of network operators and regulatory offices can have a profound impact on the growth of markets: this is the case for cellular phones, whose development has been fast in the UK but until now, very slow in France.

Globalization

The internationalization and globalization of the telecommunications equipment industry is part of a more general tendency concerning the world economy and particularly the IT sectors. As we have seen in Chapter 3, globalization is a complex process. In the present case, three factors are playing a prominent part.

The first factor which seems particularly important in telecommunications equipment is the rise of the minimum efficient scale (MES) due to rising R&D costs. The fact is well known in the case of public switching (see Table 7.1). As a consequence, the domestic market in each European country has become too small to cover these indivisible costs. In business systems and terminals, R&D costs are also growing fast, mainly because of the need for diversification and customization of products, the shortening of the life-cycle of each product generation and, more generally, the characteristics of the new technological trajectory, based on a systematic exploitation of scientific and technical advances (see Ch. 2).

A second factor is the slackening demand for some products, public switching systems in particular, in the largest industrialized countries, including France. This forces European as well as American firms to explore new markets. In Europe, for example, there are still large countries where the telecommunications infrastructure is relatively underdeveloped, i.e. Italy and Spain.

But the third and main factor concerns the general evolution of telecommunications and the use of telecommunications services. In connection with technical and institutional changes, new strategies and forms of competition, the equipment industry is turning, using the terminology of Porter (1980, 1987), from a "multi-domestic" into a "global" industry, i.e. an industry in which the position of a firm in each country is dependent on its position in other countries and where competitive norms are determined on a world scale. The characteristics of the telecommunications industry explain why globalization came relatively late. Until recently, in France and the UK particularly, firms were orientated mainly towards the domestic market. But markets have been changing very rapidly since the mid-1980s. Access to international markets is becoming a condition of survival and the whole European industry is restructuring itself through multiple mergers, acquisitions and alliances.

So, firms have to face new challenges: fast and complex technological changes, new forms of competition and relationships with users, and globalization of markets. Let us examine how they try to face these challenges.

Strategies and organization

Strategic choices

The telecommunications equipment industry is highly concentrated. In Europe, the ten largest firms account for 82% of the market (Table 7.7). Our analysis is focused on a sample of 15 firms, the main British and French telecommunications equipment firms and some representative multinationals present in the UK or France, or both (Table 7.8).

Table 7.7 The largest European tele-communications equipment producers; total market: 15.6 billion ECU in 1987.

	Market share (%)
Alcatel	28
Siemens	18
GPT (GEC–Plessey)	7
Ericsson	6
Bosch	6
Philips	6
Italtel	4
Matra	3
Sagem	2
STC	2
Telettra	1
Racal	1
APT	1
Others	15

Source: BIPE, *Le Monde Informatique*, March 1989

Facing the same challenges and constraints, the firms have various structures and strategies, reflecting their history, their capabilities and their choices. Table 7.8 gives an idea of the diversity of their situations. The differences relate mainly to three strategic axes: degree and forms of diversification, degree and modalities of internationalization, and acquisitions and alliances policy.

The positions and strategies of firms relative to diversification and specialisation are very different in two respects: the degree of diversification *within* and *outside* the telecommunications sector (or the degree of specialization in telecommunications).

A contrast exists between firms that cover the full range of telecommunications products and "niche firms" specializing in specific product or market segments. The first are the "giants" of the telecommunications business: AT&T, Alcatel, GPT, and Ericsson, all present in the public switching markets. Some firms like Motorola, STC (Northern Telecom) or Matra are in an intermediate position. Secondly, giants as well as niche firms can be specialized in telecommunications or diversified in other activities, usually in electronics with, in some cases, elements of vertical integration (AT&T, or Siemens in electronic components). Alcatel and Ericsson are good examples of the first case. Both have left some of their previous activities to concentrate on their core business (defence electronics, electronic components and consumer electronics for Alcatel; computers for Ericsson, after selling its Data Systems division to Nokia). As a result of the conditions of its creation as a telecommunications joint-venture of GEC and Plessey, GPT is in a similar position. AT&T, GEC–Plessey (Siemens) and Motorola are, on the contrary, very diversified outside telecommunications equipment, as are such niche firms as Racal, whose main activities are in defence and security electronics, or the Sagem group, which is active in defence electronics and cables. Matra is another typical example of a diversifier in different high-technology fields.

Another important aspect of strategy is diversification towards network carriers and services. As we will see later, nearly all telecommunications equipment firms diversify, more or less, towards services. But in that respect AT&T is in a specific position, being the only telecommunications firm (maybe with Northern Telecom) to offer the full range of equipment and telecommunications services. This gives it full competence in network management, the capacity to supply complete solutions and not only equipment, and as such it has a strong market position. As we will see, Alcatel's latest reorganization, like that of most large full-range product firms, aims at the same broad diversification. In that case, the strategy is

Table 7.8 The telecommunications equipment firms. (Names: [upper group] full-range product firms bold ; [middle group] hybrid firms *italics*; [lower group] niche firms roman.]

Company	Diversification viz. telecommunications equipment		Vertical integration	Diversification in services	Export (in telecom)	Globalization	Acquisitions	Alliances policy	Organization	Strategic axis	Evaluation/ remarks
	Outside	Inside									
AT & T	Strong	V. strong	Strong	V. strong	Low	Weak, increasing by alliances	Selective	Diversified strategic A. (in Europe)	"National account managers" "Strategic partner approach"	Global supply of products & services	Very competitive Penetration in Europe
ALCATEL	Weak	V. strong		Weak, increase	High (Europe)	V. strong (Europe)	V. important	Diversified, few strategic acquisitions	5 product groups, competence centres, global	Globalization by acquisitions	Main problem: internal rationalization
Ericsson	Weak, decreasing	Strong	Strong	Weak	High (Europe)	Strong		Diversified	7 vertically integrated customers area-comp. global	Back to core business	
GPT	Weak	Strong	Strong		Low	V. weak			4 product groups Matrix structure		Now subsidiary of GEC/Siemens (1989)
GEC	Strong	Intermediate	Intermediate	Weak	Low	Weak	Important	Strategic A. large joint-venture	Federation of firms	Globalization by strategic alliances	Still weak in innovation
Plessey	Intermediate	Strong, decreasing	Intermediate	Increasing pre-merger	Medium	Weak	Important	Diversified	"Business divisions", towards matrix within business	Back to core business from manufacturing to services	Unsuccessful, absorbed by GEC/Siemens (1989)
Motorola	Strong	Intermediate	Intermediate	Increasing	Medium, increasing	Strong			7 sectors Flexibility	Be world no. 1 in mobile telecom., diversified in services	
Matra communication	V. strong	Intermediate, increasing		Weak	Low	–	Important	Strategic A., large joint-venture	5 product lines Matrix management (in facilities)	Globalization through strategic alliances	Fast growth in telecommunications
STC	Strong, increasing	Intermediate	Decreasing		Medium	–	Important (ICL)	Many, no long-term	Many business divisions Matrix management	New markets (USA), realignment	Unsuccessful realignment Acquired by Northern Telecom (1990)

SAGEM–SAT	Intermediate, weak	Intermediate	Strengthening	Weak	Medium	–	Few	Diversified, careful, no joint-ventures	Product branches Form of matrix structure	Niche strategy, stay independent	Fragile
Racal	Strong, decreasing	Weak		Very strong	Low	Weak	Few	Careful	"New teams for new business"	Shift out manufacturing into services	
Case	Weak	Weak	Strong	V. weak	Medium or high	–	–	In R&D			Fragile
Orbitel	Weak	Weak	Weak	V. weak	Low	–	–	Many		Licensing More volume production	
Jeumont –Schneider (Telecom)	Strong	Weak		V. weak	Low	–	–			Efficient size by acquisitions	Unsuccessful, absorbed by Bosch (1988)

to organize a joint supply of equipment and services. Moving away from manufacturing into services (as, for example, Racal has done) is a very different strategy.

All telecommunications equipment producers seek to expand to a world scale, but following different logical paths. We can identify two basic forms of strategy on a world scale: globalization, and export on a national production basis.

Globalization can be seen as the dominant form of expansion of large multinational corporations. Global firms are organizing both their manufacturing and R&D on a world scale. Global organization differs according to the mode of spatial division of labour (see Dicken 1986). The truly global corporations are choosing a globally integrated production strategy, aiming at realizing economies of scale, more particularly in R&D, and taking advantage of different locations. That is the case of Alcatel (after the acquisition of ITT telecommunications activities in Europe), Ericsson and Motorola.

Other firms seek to develop on a world scale, while maintaining a national production basis. As a rule, the niche firms are in this situation. This strategy can have the advantage of a more consistent and flexible production system in a few key technologies. For these firms, globalization is restricted to commercial subsidiaries and some production units in countries where domestic manufacturing is a condition of entering the market (a common situation in telecommunications). The formation of alliances will be particularly important to cover R&D costs, and to penetrate certain markets such as that for cellular phones.

Some of the large full-range product firms remain far short of true globalization, such as GPT. Even AT&T is only just beginning to develop in that direction, by the way of strategic alliances, particularly in Europe, for example with Italtel in Italy and Telefonica in Spain.

Acquisitions and alliances are the main ways to achieve critical size and internationalization. Acquisitions and mergers have played a prominent part in the restructuring of the telecommunications equipment industry on a world scale since approximately 1985. The same is true for strategic alliances. Acquisitions and alliances are partly alternative and partly complementary means of achieving similar goals: to develop or acquire technologies and know-how, penetrate new markets, reach an efficient size for some activity, particularly in R&D, increase market power, and so on. Alliances can have specific targets. These are particularly important in C&C technologies, for example, seeking to impose a norm, or assure connectivity between systems. They can also be a necessity to enter

168

protected markets. All examined firms combine acquisitions and alliances, but they differ with respect to their acquisitions policy, which relates mainly to their size and financial capacity, and the way they use alliances. We can identify three main groups.

The first group views acquisitions as a means of growing to an efficient size and attaining globalization; this has been Alcatel's strategy, with the acquisition of the telecommunications activities of Thomson in 1983 and those of ITT in 1986. Siemens is another example, with a number of acquisitions in telecommunications as well as in computing.

A second group is more alliance-orientated. Strategic alliances, particularly large joint ventures, are used as the principal means for growth and diversification in telecommunications activity. GEC strategy is clearly of that type. Matra provides the best example of a strategy focusing on alliances in telecommunications, with joint ventures with Ericsson and Nokia, as well as in other sectors such as defence electronics and aerospace. Strategic alliances can also be aimed at joint acquisitions: for example, the alliance between Matra and Ericsson for the acquisition of CGCT, and between GEC and Siemens to absorb Plessey.

The specialized medium-sized firms favouring a niche strategy and independence follow a third type of strategy. They cannot afford major acquisitions, and often seem to avoid important joint ventures; Sagem is an example. However, these firms can pursue multiple alliances in different fields, particularly in R&D, usually constituted with firms similar in size and strategy.

Trajectories

On the basis of size and degree of diversification in telecommunications products, we distinguish three categories of firms: the largest, covering the full product range; the "niche" firms specializing in specific products and markets; and, as an intermediate group, some "hybrid" firms.

The large equipment suppliers: which way to globalization? The largest firms started as the main suppliers of public equipment. But the transformations of the telecommunications industry have forced them to expand rapidly in global markets. Their development has been directly linked to the conditions of restructuring in each country. In this respect, there is a marked contrast between France and the UK.

At the end of the 1970s, after a first important restructuring phase led by the DGT, the French telecommunications industry was dominated by two French groups, CIT–Alcatel and Thomson–CSF. In the 1980s Alcatel became after AT&T the second largest telecommunications equipment firm in the world. The Alcatel group today is the result of two major developments, managed by the firm itself. First, after an agreement between Thomson and CGE in 1983, the telecommunications activities of Thomson were transfered to Alcatel, making it the main French telecommunications equipment firm. Secondly, Alcatel acquired the public and private telecommunications activities of ITT in Europe (excluding the UK) in 1986. These operations transformed the company entirely, endowing it with a sophisticated, but still imperfect, digital exchange technology (System 12), and providing access to hitherto closed markets in Europe and an important production and R&D base in Europe.

Alcatel NV, created in 1987, is a truly global group, with subsidiaries in 75 countries and major facilities in 21 countries. In 1988, 60% of Alcatel's output was produced outside France and 69% of that output was destined for non-French customers, mainly in other European countries. Alcatel NV is in fact a European group, keeping its accounts in ECU. On the whole, the strategy of Alcatel can be characterized by: globalization by acquisitions on a European base, and concentration in telecommunications with a full product-range strategy. The main challenge is now, for Alcatel, to rationalize its production structure at a world (or at least European) scale, to make product lines coherent and to integrate different cultures. The latter is probably the most difficult task.

The British story is quite different. At the end of the 1970s, the British public switching market was still shared by three firms: GEC and Plessey, jointly developing System X, and STC. The decision of British Telecom to have a single UK supplier and to cede 30% of the digital exchange market to a foreign firm (Ericsson) initiated the restructuring of the industry (see Morgan 1989). STC was ejected from the market after a transition period and Plessey became the main contractor. In 1986 GEC tried, without success, to enforce a merger with Plessey. But bringing together the telecommunications capacities of the two firms was clearly necessary to face world competition in telecommunications. In 1988, the telecommunications divisions of GEC and Plessey were unified in GPT (GEC–Plessey Telecommunications), a common subsidiary of the two companies. The new firm became the third telecommunications equipment company in Europe, and the seventh (Morgan 1989) in the world in terms of sales. But for production as well as sales, it remained largely UK-based.

The evolution of British markets, strongly influenced by BT strategy, has led the main UK telecommunications firms to redefine their strategies. GEC, a large diversified company (in fact more a federation of firms than a real industrial group) is not very dependent on its telecommunications business (9% of its turnover in 1988). But in recent years, it has engaged in a series of strategic alliances, in the form of joint ventures, across different business interests, directed to entering global markets. Among these alliances was one with Siemens for defence electronics and telecommunications. Plessey was also a diversified company, in defence and electronics components, but less than GEC; and it was much more dependent on telecommunications markets. Also more technology-orientated and more dynamic, it was looking to focus on this core business and to move into services, mainly by acquisitions.

After long resistance by Plessey, the GEC–Siemens takeover bid succeeded at the end of 1989, making of GPT a joint venture of GEC (60%) and Siemens (40%). The fate of GPT remains uncertain, but it seems likely that Siemens will be the leading firm for telecommunications. Thus, the globalization of the British telecommunications equipment industry will finally be carried out, but by subordination to the Siemens empire.

The niche firms The specialized firms, medium-sized or large, are still important in the telecommunications industry. Some of them are specialized strictly in telecommunications (Case, Orbitel), others are diversified in several high-technology fields, very often in defence electronics (Racal, Sagem and Sat). The niche firms are trying to take advantage of high technical expertise in some specific fields, and to exploit them in different markets. This is the case of Racal (cellular radios), Case (equipment to link computers over telephone lines: modems, multiplexers for which Case is the European leader, data switches, etc), and Orbitel (equipment for digital cellular networks). The Sagem–Sat group is more diversified but pursues an explicit niche strategy, in telecommunications as well as in its other activities (focusing on telex, telecopy and teletex for Sagem; modems, optical transmission and radio links for Sat).

The condition of success for niche firms is to keep a leading position in their speciality and to expand in international markets. For this purpose, they must have a consistent and flexible production and R&D system on a national base. By a strict specialization combined with R&D and commercial alliances, they hope to face the problem of rising R&D and marketing expenditures. The growth of some of these firms has been promoted partly by access to protected public markets, especially in France, and by the rapid growth of new markets, in particular for mobile phones in the UK. But, because of growing competition in many of the markets in which they are present, and the globalization of the large

171

equipment suppliers, future growth will probably be difficult. These firms follow different strategic routes.

Sagem is expanding by using its technological and commercial capabilities to develop new products, such as telefax and multi-purpose terminals, as well as new activities such as automotive electronics. In reaction to the relative decline of public demand it has also developed mass-production products for business users. Racal, wishing to diversify out of defence, is expanding in services and shifting out of manufacturing. This last tendency, which was also at work in Plessey, seems more distinctive of British than of French firms.

The hybrid firms Some firms are situated between the two extreme forms. Motorola, a large diversified US electronics firm, has a typical global organization, with major facilities in 18 countries, including France, Germany and the UK. One of its main aims is to be the world leader in a single technology, namely cellular telephones.

STC and Matra are firms in transition, in almost opposite directions. STC, excluded by BT from the digital exchange market, had to define a new strategy. By the acquisition of ICL in 1984, it chose to diversify into computers and to expand on global markets for some telecommunications products and services (multiplexers and software) on the basis of a strong R&D capacity. STC was involved in many alliances with, in particular, the objective to expand operations in Europe, but without long-term commitments. Again, international expansion and penetration of the "solutions business" were the key market objectives. But, in some respects, STC was too orientated to niche strategies. Its recent (1990) acquisition by switchmaker Northern Telecom will give STC a presence in the switching mainstream. Moreover, its prior sale of ICL to Fujitsu (also 1990) secures that company's presence in mainstream, large-scale computing.

Matra, diversified in different high-technology sectors, has grown fast in telecommunications during the past five years. One dimension of the restructuring of the French telecommunications industry during the 1980s has been the transfer of the CGCT assets to Matra and Ericsson, in two stages. First, in 1986, Matra acquired the private switching and business systems activities of CGCT, nationalized in 1982. Secondly, the creation of MET in 1987 (a joint venture between Matra and Ericsson assured the public switching activities of CGCT) MET becoming a second supplier to DGT. It transforms Matra Communication, the telecommunications subsidiary of the Matra group, into the second French telecommunications firm. Matra has been expanding its operations in the main growing telecommunications markets (public switching, PABX, mobile phone, teletext, telecopy), and focuses on high technological standards. Strategic alliances, in particular large joint ventures, are central in the Matra policy and its strategy of globalization.

Overall, there are likely to be two basic possible strategies: globalization, in terms of both products and geographical location, and the niche strategy, which is often imposed by sectoral evolution rather than being clearly chosen by the firm. At the same time, firms may develop very diverse strategies in response to cultural and institutional factors. These can have a profound impact on their development strategies.

Organizational dynamics: between globalization and dynamic flexibility
The transformation of the technological regime and the competition norms are leading to new organizational forms. Global organization and the search for dynamic flexibility are the most important tendencies.

Global organization   For the large telecommunications equipment suppliers, penetration of international markets implies a complete globalization: the organization of their activities on a world scale. The internationalization of production can take different forms (Dicken 1986, Savary 1989). According to the *multinational form*, production remains organized on a national basis, without geographical specialization. In the *global form*, by contrast, we find a globally integrated production strategy with a specialization mainly of national manufacturing and R&D units. Global organization can be *vertical* or *horizontal*. In the first case, the production units in one country are specialized in a production stage; in the second case, the production units are specialized in a product type, or a technology, for the world market, or for a regional market of several countries (Europe, for example).

According to Porter (1985), horizontal globalization is the predominant organizational form for multinationals. This is borne out for firms in the telecommunications industry. Two factors can explain this: one is the progressive unification of markets, providing the opportunity to exploit scale and scope economies; the other is the diversification of the large equipment firms, linked to the multiplication of telecommunications products and systems, as well as diversification outside telecommunications for some firms. For large diversified firms, horizontal organization seems to be the most efficient form of production rationalization.

At the same time, there can be technical and market specificities and political constraints, which means that some production units, and in some cases development units, have to be specialized for the local market. This is certainly the case for telecommunications equipment.

Alcatel and Motorola are good examples of firms with global organization. Motorola has, for example, 17 research centres in Europe, each specialized in one domain. The presence of research units in Europe is partly explained by the need to develop products specifically for Europe. The case of Alcatel is very significant in terms of a global organizational strategy. Following the acquisition of ITT's telecommunications activities, the main problem has been the rationalization of the group on a European scale, implying the development of a global organization, in two different ways. First is the creation, at the headquarters level, of five "product groups" in charge of development, manufacturing and marketing strategies for each group. This organizational form is not unusual in telecommunications firms. In Alcatel's case it must ensure the co-ordination and rationalization of subsidiaries in different countries. Secondly, the regrouping of R&D units under the same direction. The very high costs of R&D make rationalization of that activity particularly important. Some subsidiaries have been chosen to run specific research fields, thus becoming "competence centres", a typical form of R&D organization in transnational corporations. That type of organization is also found in Ericsson. A competence centre is in the first place an R&D centre specialized in one product line. This can be connected to manufacturing facilities in the same location or in different locations in different countries.

Globally organized firms frequently keep R&D, for example, in specific locations (often in their country of origin and a few outposts). Thereafter, as functions become less research-intensive, the choice of location of an activity is less constrained. Local manufacturing capabilities have been established by Alcatel in several countries for local markets, in order to maintain strong relationships with clients and to implement custom-designed manufacturing programmes, in particular for public networks systems in developing countries. On the whole, the production system of the global firm takes the form of an integrated international network of R&D, manufacturing and commercial facilities with an internal division of labour, capable of supplying a large range of complex products and systems.

In search of dynamic flexibility All firms involved in global or niche strategies have to face the constraints imposed by what we called the new regime of "dynamic flexibility": the capacity to develop new products and to respond rapidly to the specific needs of different users. The shortening of product life cycles can be quite spectacular for some telecommunications products: the first telex had a life-time of 20 years; for modern terminals it can be less than 16 months. This tendency is also connected to growing competition on the basis of product quality and performance. In the case of telecommunications equipment, the privatization of markets and the increasing importance of business and consumer products have also had important consequences for the organization of firms.

The general tendency is towards a greater integration of R&D with manufacturing and marketing. This is a necessary to accelerate the development and realization of new products, and to adapt the design of products and systems to market needs and manufacturing constraints. Integration is achieved in several ways. Usually, firms are organized in product groups, covering development, manufacturing and marketing. A transverse structure can ensure synergies between product groups and between research and production. The divisions are usually organized by client sectors or business. However, there may also be some form of matrix organization, partly explained by the complexity of products and systems and the need to supply "global solutions" combining different products and technologies.

The use of flexible manufacturing systems and the generalization of CAD/CAM systems facilitates direct links between development teams and manufacturing facilities. Computerized systems are also used to manage large product ranges and to obtain economies of scale by standardization of components and modular conception of products. Flexibility in manufacturing can also be attained by a flexible specialization of manufacturing units: each facility is specialized in a category of products, but can be used for other products if demand fluctuates. In some firms neither manufacturing units nor R&D laboratories are tied to a specific product division. Teams can be made responsible for programmes, projects, new products or new business, from conception to industrialization and commercialization. For example, at Racal we find the principle of "new teams for new business", or at Sagem a new structure called "the three partners team" consisting of a product leader, a research engineer and a manufacturing engineer jointly starts a new project. It can be seen as a form of flexible matrix organization.

In all firms, even the more technology-orientated ones, the commercial function is seen as crucial. It reflects the rapid change of telecommunications markets and demand structures. In spite of the need for strong integration between R&D, manufacturing and marketing, separate commercial structures often occur. In all French firms we studied, we found the same choice, at least for business systems: a separation between distribution and production by the creation of distribution subsidiaries. The development of new private markets is the first reason for this choice: it seems the best way to create a commercial "culture" in firms previously orientated towards public demand. It can also be a means to clarify the relationships with France Télécom. And it provides a way for optimal adaptation to user needs: the distribution subsidiary does more than mere marketing by the customization of products and the supply of client "solutions".

Last but not least, the tendency to move from products to integrated systems and solutions supply also implies new organizational structures. This process is associated with the diversification to services, as we will see in the next chapter. It concerns in particular the supply of complete customized public or private networks. Alcatel recently modified the definition of its product groups by the creation of a "Network Systems Group", taking the place of the "Public Network Group". Exactly the same structure exists at GPT. It regroups public switching, public and private networks (which are becoming very similar) and cable transmission systems (including fibre-optical systems). The logic of that reorganization is clearly explained by the chairman of the new group:

"a networks or services approach is from now on preferable to a simple products approach" (*Informatique*, 26 January 1990)

On the whole, the most characteristic feature of organizational dynamics seems to be the need for permanent reorganization. The problem is not to find the best structure, but to be able to adapt the organization quickly to technical and market changes, to attain organizational flexibility.

Conclusions

The transformation of the telecommunications industry during the 1980s has been extremely rapid. Restructuring will continue with the diffusion of new systems (ISDN in particular) and the gradual liberalization and unification of European markets.

The technological factor has been the starting point of this evolutionary process, radically altering the economics of telecommunications. But the recent story of the industry, not least in France and the UK, shows the importance and autonomy of institutional factors, which are the main cause of the divergent trajectories of the equipment industry in the two countries.

The rôle of cultural factors should also be noted. The difference between the cultures of GEC (a large finance-orientated and traditionally managed company) and that of Plessey (more dynamic, technology-orientated and wishing to stay independent) helps to explain the difficulties in restructuring the UK telecommunications industry. Cultural differences are also one of the reasons why, in spite of the technological convergence, the diversification of telecommunications firms towards the computer industry has been, until now, unimportant and often unsuccessful.

Overall, increasing concentration and domination by a few global, integrated companies is the main tendency. Paradoxically, deregulation and liberalization of markets are promoting such developments (Roobeck 1988). The diversification of products and markets, and the constitution of international networks of alliances, can allow niche firms to survive and expand. But rising R&D expenditure, increasing intensity of competition on European markets and the raids of large companies in search of new acquisitions, will put most of these firms in a difficult position.

With Alcatel becoming a world leader in the sector, the restructuring of the French industry has been more successful than the British. But, after a long period of state protection, it remains to be seen if Alcatel will be able to adapt to the new competitive configuration of the world industry. The poor results reported up to 1990 in private systems and terminals are a significant indication of the problems Alcatel has to face.

CHAPTER EIGHT
Services:
the bridge between
computing and communications

Frank Moulaert

The world of services related to computing and communications is even more complex than that of computing and communications equipment manufacture. This complexity is mainly linked to three groups of factors: (a) by their nature, professional services, including many C&C services, are designed to satisfy user needs in a customized way; (b) C&C services assure the bridge between C&C equipment, information suppliers and information users; (c) C&C services are badly needed to solve problems of compatibility, adaptability and connectivity between C&C systems. As a consequence, the technical content and the use value of C&C services can be very diverse and sophisticated. Production and delivery of these services can not therefore be analysed through simple extrapolation of knowledge about the production and distribution of C&C equipment and goods.

C&C services can be provisionally grouped under several labels: (a) software services, (b) network services, (c) consultancy and systems integration, and (d) facility management. These groups partly overlap, and also demand more detailed analysis to be really understood. That is particularly the case for network services, which include electronic data interchange (EDI), electronic mail services (EMS), databank retrieval, and on-line information services available in videotext form (Department of Trade and Industry 1988, OECD 1988b).

Even more complex than the world of C&C services is the world of their suppliers. Later in the chapter, we will present an exhaustive overview of suppliers and the services they provide; here, we only provide an idea of the complexity of supply structures. Direct suppliers of C&C services range from C&C hardware producers, through systems and software houses, to consultancy and accountancy firms, network managers, telecommunications service providers, databank managers, and user-firms, who often become important suppliers. In this sense, also making important distinctions between the different services which are provided, we are faced with a complex ensemble of new activities, which may yet become a distinct service sector.

Let us look at two examples. Information technology consultancy (ITC) – basically corresponding to "the intellectual activities preparing and accompanying the installation of information technology systems in user organizations" (Moulaert et al. 1988) – is a service provided by a wide spectrum of suppliers: IT consultants with an accountancy origin (such as Arthur Andersen); engineering consultants with a "Taylorist" origin (such as Bressand in France or Berenschot in The Netherlands); systems houses which have become major IT consultants (such as the Cap Sema Group, and Cap Sesa, through the merger of Sesa and Cap Gemini); hardware producers (such as IBM, AT&T who developed a significant systems integration activity[1]; and different types of high-level professional service supply. To complicate matters even more, not all "consultants" offer the full range of ITC or related services. Some very specialized IT consultants will do only strategic and functional analysis as well as the design of IT systems. Others will provide the full systems integration job, including analysis, but also hardware selection, software selection or construction, operationalization, training of end-users, etc. Still others will also offer facility management and network services.

Another example concerns the value-added network services. Ranging from electronic mail, through electronic data interchange, to information retrieval and transaction services, these may be provided by very different suppliers. The fact that in many countries public telephone lines remain a major carrier of these networks assigns an important rôle to public telephone companies, without compromising the rôles of other parties. Such other parties can be hardware producers such as IBM, with its Business Network Services; Siemens through Vascom AG; databank managers such as GTE Telenet (see Th. Ribault in ERMES 1988); facility managers such as EDS, now part of General Motors Inc.; or original network integrators such as Istel, now acquired by AT&T (*Communications Week International* 1989).

Even more than is the case in manufacturing, C&C services are provided by suppliers with very different sectoral origins. The integration of the service market is more of a reality than that of computing and telecommunications technologies themselves. In fact, in manufacturing, only the really global firms are able to combine successful production and market strategies for both computing and communications equipment. In services, more hybrid and medium-sized firms operate successfully in integrating different elements of service solution.

Statistical appraisal of C&C services

How can one grasp in statistical terms a set of activities whose use values constantly change, which are supplied by myriad different producers, and which are the subject of transactions following nationally quite different technical and trade norms?

Looking at the list of different suppliers, we can conclude that the available *international* data (comparable data for different countries) are sector- and market-biased. By sector-biased, we mean that markets are defined according to the output of their main suppliers. For example, the estimates of the computer-service market are based on extrapolations of the turnover of major systems houses and hardware producers with a majority of service activities. They exclude the very important part played by accountancy firms, engineering consultants, etc. And they over-estimate the rôle of the major suppliers, by taking into account their turnover realized in non-C&C service markets. By market-biased, we mean that only very specific, easily discernible, product markets are identified. Telecommunications services in general can be statistically evaluated, but particular services (such as enhanced network services) are much more difficult to identify and measure. These are provided by so many different types of suppliers, are included in larger service packages (e.g. network services in facility management) and their nature evolves so rapidly, that clearcut distinction between service groups, and therefore of markets, becomes difficult.

For many of the C&C services, there are reasonably good *national* sources. However, not all services are covered, and international comparison is impossible. The availability of national data for France and Great Britain can be illustrated by looking at turnover in the different sectors.

For France, interesting and detailed data are provided for computing services, both sector- and market-wise (Moulaert et al. 1990). These data are provided by the French national statistical institute, INSEE, and by the Ministry of Industry's statistics (in co-operation with SYNTEC, the French professional organization of systems and software houses)(Table 8.1).

In the UK, official data on high-technology consultancy are blurred by the inclusion of other activities in SIC MLH8394, which is defined as comprising data-processing services, consultancy, recruitment and training, and software production. Moreover, this data does not include consultancy services in ITC offered by those outside the software sector itself. The UK industry body, the Computer Services Association (CSA), gives the information on revenue distribution (Table 8.2). Again, however, only part of the whole ITC consultancy business is contained within the CSA, and there is little agreement over the definition of terms.

Table 8.1 High-technology consultancy in relation to the professional services sector in France (NAP 77, 1986).

Subsector (NAP-3000 level)	Employment	Turnover	Value added	Investment
77013 Special and technical studies in manufacturing	44,198	26,875	11,135	835
77031 Management consultancy with respect to HT applications	12,839	5,940	3,669	166
77035 Computer and information technology consultancy non-customized software	37,725	16,865	9,807	889
77038 Systems engineering	6,212	3,688	1,907	147

Source: Moulaert et al. (1990).

Table 8.2 Revenues for CSA member companies, 1988/89 (%).

	%		%
Custom software	18	Education and training	4
Software products	12	Processing	7
Systems integration	18	VANs	2
Consultancy	19	Databases	2
Facilitator management	7	Recruitment staff	4
Transaction processing management	4	Other	4

Source: CSA annual report (1988).

181

From the discussion above, it becomes clear that statistical analysis can hardly be the unique tool to analyse the development of C&C sectors and markets. Continuous sectoral innovations, insufficient coverage of all products in the market, lack of statistical detail and comparability, not to mention accounting problems, affect the reliability of the national statistical apparatus in a negative way[2] International sources use a coherent statistical approach to the different national sectoral settings. While they are able to deal with comparability and consistency problems among countries, they have the disadvantage of being samples, often composed of leading firms in the market, with strong affiliations to established professional associations. This might lead to a somewhat rosy view of the C&C markets. Let us look at some international comparable data on the development of the information technology software and services market (Tables 8.3 & 4).

Table 8.3 Software and services market forecast, Great Britain, 1988–93 (£ million).

Market segment	1987		1988	1988–93	1993
	£	%	£	AGGR	£
Processing services	517	13.82	560	7%	780
Network services	165	4.41	220	28%	740
Software services	970	25.94	1,200	22%	3,200
Professional services	1,200	32.08	1,470	20%	3,720
Systems integration	205	5.48	255	24%	760
Turnkey systems	680	18.18	810	17%	1,730
TOTAL (rounded)	3,737	100.00	4,515	19%	10,930

Source: Input.

Table 8.4 Software and services market forecast, France, 1988–93 (FF million).

Market segment	1987		1988	1988–93	1993
	FF	%	FF	AGGR	FF
Processing services	10,500	19.26	11,010	4%	13,200
Network services	1,380	2.53	1,900	26%	6,140
Software services	12,880	23.63	19,670	22%	41,720
Professional services	19,120	35.08	23,360	20%	58,180
Systems integration	1,735	3.18	2,190	28%	7,260
Turnkey systems	9,195	16.87	10,850	17%	23,650
TOTAL (rounded)	54,810	100.00	64,980	18%	150,150

Source: Input. (AAGR: average annual growth rate)

These data show tremendous growth rates for all services, with the exception of processing services (Moulaert et al. 1990). Network and professional services are more developed in the UK, software and processing services more in France. Clearly, network services are still in a very early growth stage.

C&C service use-values and market integration

Even if statistical data on C&C services had been of a higher quality, statistical analysis would still be insufficient to grasp the qualitative dynamics of changing C&C service markets and supply structures. The nature of C&C services is very different, depending on their function in the information systems of the organizations they serve. And, as noted in the introduction, similar supplies can be delivered by very different suppliers.

C&C service use-values

The analysis of an information system helps to understand the different kinds of C&C services. The information-system concept encompasses one element and two actions: information, (tele)communications and processing. The former needs the latter two to have use-value; the latter two depend on the former to be able to work. The system stocks, transmits and processes information. To operate, it uses hardware, systems and application software, and "orgware", which assures the correspondence between the information system and the organization (Moulaert et al. 1991).

In purely technical terms, a distinction can be made between the following types of C&C services necessary for:

(a) construction of data-processing systems,
(b) construction of communication systems,
(c) processing of data,
(d) transmission of data,
(e) [training, maintenance and management] assurance of system construction and operation.

From this technical point of view, the classification of C&C services looks relatively straightforward (Fig. 8.1). However, as argued elsewhere, a purely technical approach to C&C services is inadequate. Information systems can be considered as the "spine" of the organizations implementing them. As such they are affected by the internal and external dynamics of

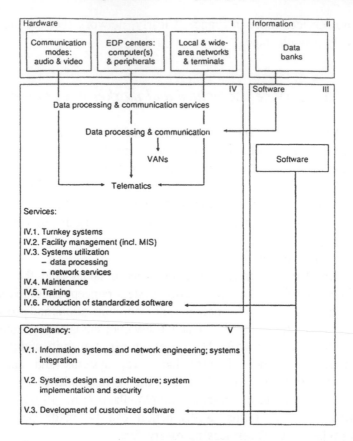

Figure 8.1 IT consultancy and services (*Source:* Moulaert et al. 1990)

these organizations. But, at the same time, organizations use IT systems in a strategic or instrumental way to achieve their objectives.

This very observation makes a sound classification of C&C services from the point of view of users' needs practically impossible. Take the example of enhanced network services. Users will want to be hooked up to networks which assure access to the databanks needed for their business operations. Of course, users with different needs can use the same electronic mail systems or EDI carrier. Moreover, their application can employ the same type of microprocessors, with only the software and graphics being different. Still, the needs for customized data-processing and transmission are so different, and the possibilities to satisfy them so many, that the systems integration market flourishes along very diverse, hard to classify, technical and social trajectories.

Of course, that should not leave us with the impression that C&C services occur in a world of unconstrained synergies. The integration of technological C&C systems, network carriers, data collection, processing and transmission is constrained in many respects. The quest for a general open architecture, allowing communication and compatible file transfers between different operating systems offering specific applications, is far from over. Connectivity between different types of hardware is hampered by enterprise cultures of leasing firms, market niche strategies, and also private and public regulation practices. Moreover, given the uncertainty about future technological developments, adaptability of existing to new systems is not automatic (DOS versus OS/2 operating systems, for example).

Synergies at the level of integration of existing technical and applications assets will probably also become more limited by increasing standardization of basic technology, operating systems and communication environments. The market pressure towards standardized, commodified elements applicable in very different systems is significant. High development costs and the pressure to realize economies of scale, while respecting the variety of needs, are at the root of that pressure. That is why there is a tendency to move the concern for customizing from the level of hardware and systems production to the level of applications. The spread of user-friendly fourth-generation programming languages goes in that very direction.

Diversification of products and integration of markets
Diversification of C&C services can be explained by different factors, such as increased technological possibilities, more flexibility to customize products, and the growing opportunies to integrate different equipment and service products in "complex products" (ERMES 1988) or "global solutions" (Moulaert 1990).

Integration of markets can be explained by, on the demand side, the changing demand to meet more customized, complex and intellectualized needs and, on the supply side, the availability of different assets necessary to offer "global solutions", with suppliers coming from very different sectoral origins. These suppliers try either to acquire missing capacities, or to assess them through cooperation with "quasi colleagues" (peer-group "review").

On the demand side, the development of the demand (or consumption norm) among leading clients sheds a clear light on what is meant by a global solution. Leading clients have a tendency to look at their information and communication problems in a structural way. Information and communication functions belong to the spine of their business organization. As such, they cannot be isolated from other functional dynamics which they serve or on which they rely. This means that putting in place or restructuring an information system requires strategic and functional analysis of all business activities concerned. A global approach is needed and it involves all services – analytical studies of business functions and IT systems, "orgware", systems and application software, training, quality control, databank retrieval, etc.– and equipment – central processors, workstations, lines and switches, etc. – necessary to assure the optimal match between business system and information system.

On the supply side, there is an acute awareness of the growing possibilities of, and indeed the need to offer, global solutions to clients. Both technical and organizational possibilities are available. They are developed in-house or made accessible through co-operation with specialized firms. Market pressure to do so is very high: profit margins on hardware and traditional software products are narrow, professional services (systems integration, training in information methodology) and high-value-added network services offer better prospects for profitability. Potential loss of market share is compensated by substantial engagement in complex service or global-solution strategies. In these services, problem-solving from the client's point of view is a major marketing principle. It is at the basis of a close integration of R&D, "production" and marketing activities within C&C service supply organizations. In this way, functional integration belongs to the very core of professional service activity (Moulaert et al. 1990).

Following this logic, many large C&C firms – hardware as well as service providers – build up different C&C service activities to become global C&C service firms. Table 8.5 illustrates how the major firms in the French–British sample we studied have diversified towards different C&C services, and how they have succeeded in integrating these services in a global supply package. To give the picture greater comparative value, we have added some leading consultants and software houses not included in the sample.

In general, we can say that only the *global* C&C hardware producers have successfully operated on different C&C service markets. And, while their position on the systems integration and consultancy market is qualitatively and often quantitatively not comparable to that of the largest consultancy firms, they seem to have done a better job in developing their high-value-added network services than have most of the systems houses or consultancy firms.

Table 8.5 Involvement of suppliers in C&C service provision.

	Main activity	Reach of service market	Software	ITC/ Systems integration	Network carrier	HVA Network services	Mobile telecoms networks	Facility management	Network integrator
AT & T	Telecom service & exchange equipment	Global	Yes UNIX	Yes	Yes	Yes	?	Yes	Yes
HW	Computer hardware	Global	Yes network & applications	Yes	No	Yes	No	No	Yes
Case	Network hardware	Global (but thin)	Yes esp. open systems	No	No	Yes	Yes (Case Radiotext)	No	Yes equipment more than services
GEC	Computer software	Global	Yes UNIX	Yes	No	Yes	No	No	Yes but not always
Ericsson	Telecom equipment	None	Yes	Yes some	Yes (only in Argentina)	No	No	No	No
Hewlett Packard	Computer hardware & instruments	None	Yes HP precision architecture	Yes	No	No	No	No	No
Fujitsu	Computer hardware & telecoms & components	Japan & Japanese MNCs	Yes but limited	Yes	No	Yes "Fenics"	No	No	No
BM	Computer hardware	Global	Yes	Yes	No	Yes	No	Yes	Yes
GPT	Telecom exchange equipment	Limited	Yes	No	No	Yes video conferences	No	No project management	Yes limited
Racal	Electronics (esp. military)	Europe	Not much	No	No	Yes e.g. mobile data	No	Yes very important	Yes in only
Motorola	Electronics hardware	Global	Yes	Yes very important	No	Yes	Yes	Yes (?)	Yes
TC/ICL	Telecom & computer	Limited	Yes	Yes	No	Yes some	No (?)	No	Yes
Alcatel NV	Telecom equipment	Global	Yes	Yes	No	?	?	Yes	Yes
agem	Telecom defence	Europe	Yes	?	No	?	?	No	No
AT	Telecom equipment defence	Europe	Yes	No	No	?	?	Yes	No
Matra	Telecom defence space autom	Europe	Yes	?	No?	?	?	Yes	?
rthur nderson	Accountancy management consult.	Global	Yes	Yes	No	Yes?	Yes	Yes	Yes (consult. only)
ap. SESA	Systems house	Global	Yes	Yes	No	Yes?	Yes	No	Yes?
ogica	Software house	Global	Yes	Yes	No	Yes	No	Yes	Yes
EMA Group	Software	Europe	Yes	Yes	No	No?	Yes	No	No?

Market integration and firm organization

What are the organizational and production strategies of C&C service suppliers *vis-à-vis* the growing integration of markets? To answer this question, three aspects of organization are examined: business organization, production process, and spatial organization.

Sectoral dynamics and business organization

We found that, for firms to be able to supply "complex products" or "global solutions", their organization has to integrate marketing considerations in its different functions as well as in their co-ordination. That means that service suppliers should develop a flexible response system able to deal with diverse types of demand. For example, hardware producers who become systems integrators are forced to work according to a dynamic matrix organizational structure: co-operation is necessary between product and business lines as well as market sectors. Systems-integration project groups are usually constituted with personnel having different technological, product and client-sector backgrounds, coming from different departments and business lines in the firm. The composition of project teams is variable according to the nature of the project. This often requires breaking with Fordist organizational models, borrowed from manufacturing, to give way to a more *ad hoc* style of organization (Mintzberg 1988).

Only "hybrid" and "global" firms proved to be successful in developing flexible supply capacities for a wide spectrum of services, facilitating good systems integration. Smaller and niche firms have limited chances of becoming systems integrators. If they are entering the systems integration market at all, they usually do so as members of a consortium, in joint ventures, or as subcontractors for a systems-integration project manager.

Very interesting forms of matrix co-operation in systems integration can be found within consultancies having an organization consultancy or accountancy background (Daniels 1988). They are the "high-level consultancy" organizations who undertake strategic analysis of organizations, functional analysis, and design of their information systems. Often, they are also involved in construction and management of information systems. They usually have a high degree of methodological sophistication. Moulaert et al. (1991) found that, within these consultancy organizations, there are major cleavages between project specialists, methodologists and technology specialists. Project specialists are usually trained from the point of view of a client sector (manufacturing, distribution, banking, etc.) and/or a client business function (stock administration, logistics, accountancy, sales & marketing, etc.). Methodologists are trained in the use and development of tools for analysis and design. Technology specialists follow the hardware and packaged software markets. Project specialists are assigned to projects and will follow them exclusively.

The constitution of teams and functional units within supply firms, and the interaction between them, are dynamic processes. A computer-services corporate-strategy manager observed: "We notice strong professional interaction between functional units and teams. Units and teams are often organized by client sector, client business functions and methodological or technological specialization. Nevertheless, consultancy business shows a remarkably high professional flexibility. Teams can be switched according to functional requirements of large clients and projects. Know-how is transversely combined; jobs are redefined and professional expertise is generated in relation to changing market and research needs."

Institutional forms of intra- and inter-firm co-operation develop rapidly. Power relations play their part. Hardware producers successfully try to remain in charge of large-scale systems-integration projects. Software production is quite often subcontracted to medium-size or smaller software houses. For strategic analysis, co-operation with consultancy firms is sought. Still, consultants seek to maintain and enhance their competitive advantage in strategic analysis through their client-sector experience, to keep co-operative relationships in power balance. Therefore, hardware producers – at least those we interviewed – seek to establish more stable alliances with consultants, who offer them much more professional complementarity than most software houses can do.

In sum, we observe that the development of internal service capacity in hardware-producing organizations accompanies the development of different types of occasional or more stable working relationships with consultants, software houses and other specialized service suppliers. Large consultancy organizations elaborate a more complete service supply structure; at the same time, they develop both their technology capacity and privileged, albeit often informal (non-institutionalized), links with leading hardware producers. Large software houses seek to establish their consultancy capacity through joint ventures or mergers.

C&C services: their production process
In the previous section, we saw that systems integration demands a provider-organization with a dynamic approach and a matrix-like

management structure. This general observation holds for most of the services provided, regardless of the institutional form of the supply structure, whether a single firm, a consortium of firms, an hierarchical customer-supplier network, or an occasional delivery network. In this section, interest in the institutional forms of the supply structure is only implicit, the emphasis being much more on the production and delivery process for C&C services. Since, in the social science literature, little is known about the production and delivery systems, especially for high-value-added C&C services, we believe it is useful to provide some detail on these systems, by way of concrete examples. We will do so for information technology consultancy, software production, and high-value-added network services.

Before this, however, it is important to understand the rôle of the users in the production of the C&C services. These services are, of course, meant for a market. That means that they should potentially satisfy needs. However, as with most knowledge-intensive services, they are quite often designed and produced with the effective participation of the users. Of course, the degree of commodification is quite different among the three services. Packaged software responds to specific market needs, can be improved in reaction to revealed user-preferences (through user groups, market inquiries, etc.), but is produced in close co-operation with individual users only by exception. Customized software, in contrast, is designed and developed following common specifications by user and producer. For value-added network systems, the picture is quite uneven. Leading users will certainly co-determine the nature of transmission formats, the structure of databanks, and the technical and usage features of the terminals. Networks shared by a significant number of comparable users will be designed, constructed and managed through user groups and suppliers. Information technology consultancy, finally, is perhaps the most customized of all: it seeks to develop or restructure an IT system acccording to the specific needs of the client organization, taking into account its business objectives and strategies, its functional division of labour, as well as its institutional dynamics (in which non-rational as well as rational factors play a part). Let us now look in some detail at the production system for each of the three services. The presentation here is necessarily simplified.

The information-technology consultancy process, and the interaction between consultant and client, can be represented in the form of a chain of production and interaction stages:

(1)   initialization of ITC job
(2)   strategic analysis
(3)   functional analysis
(4)   design and architecture, which imply:
       4.1 identification of functions in information system
       4.2 particularization and geography of functions
(5)   hardware and software selection (and eventually production)
(6)   implementation of the IT system.

In the *initialization* stage, client and consultant define the broad terms of the consultancy job. This might need a number of discussion rounds between the consultancy organization and general or electronic data-processing (EDP) management in the client organization. Sometimes, a feasibility study is needed. *Strategic analysis* results in a strategic business plan or its revision. This plan includes an information strategy plan. *Functional analysis* concerns different business functions and their information systems. If the consultant uses a strategic approach, functional analysis affects all functions and business areas involved in the business plan. But very often, analysis is limited to one function (marketing, stock control, accountancy), one business area, or is only operational rather than strategic in nature. Sometimes, functional analysis is divided into a normative trajectory (processes and flows as they correspond to the strategic plan) and a descriptive trajectory (processes and data flows as they exist in the client organization). In the *design and architecture phase*, a distinction should be made between the macro- and the micro-level. At the macro-level, prototypes of general information functions and flows are generated and evaluated in view of the strategic targets. After agreement has been reached on the macro-functions of the information system, micro-functions can be defined. Selected and eventually specially produced software and hardware are implemented in the information system (see software production, below).

In addition to this chain of phases, there are services which are recurrent or even concurrent with other stages: quality control during the process and between the stages; training; and maintenance of different functions of the IT system. There are important differences in style, methodology and philosophy among consultants. Sectoral origin, business culture, client sectors, national roots, and research on methodology are important in this respect (Moulaert et al. 1990).

Methodological sophistication varies considerably: the use of information systems and high-value-added services networks, the use of data models and graphics in analysis and planning tools, etc., are unequally spread among consultants. The most advanced consultants manage to "computer integrate" the different stages in the consultancy process ("computer-integrated consultancy"). Table 8.6 provides a survey of advanced consultancy tools. Methodologies differ strongly according to the consultants' philosophy of interaction with the clients, their professional and sectoral origin and background, as well as their research capacity (Moulaert et al. 1990).

Table 8.6 Stages in the ITC production–interaction chain, and tools used.

Stages	Tools
1 Initialization	1 Interviewing – orderbook – feasibility studies – documentation
2 Strategic analysis	2 Analysis and planning tools
3 Functional analysis	3 Joint application development
4 Design and architecture	4 Software development: functions & flaws
5 Software development	5 Prototyping Application generating environment Code generators
6 Hardware and software selection	6 Criteria analysis (databanks)
7 Implementation	7 System solution (4th generation language)

Source: Moulaert et al. (1990).

The development of software corresponds to stage 5 in Table 8.6. From the point of view of software production, it is obvious that analysis and design need to precede proper development. Such analysis and design can be done by a consultant who passes on specifications to a software house or a software department in a larger service or hardware firm. In any case, the specification of user needs will be quite different for customized software development than for the production of standard packages. Let us look at customized software development only. Figure 8.2 summarizes the software development process. Although the scheme is strongly inspired by Logica, its general labels are certainly valid for the different software development tasks as performed by other software houses or departments of hardware producers and consultancy organizations. Basically, a distinction should be made between the software *production* trajectory and the *quality* trajectory.

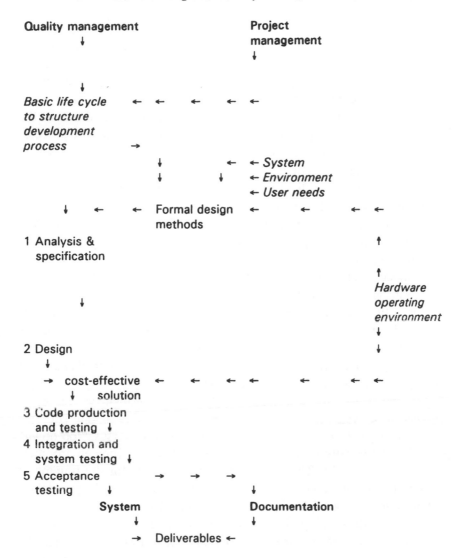

Figure 8.2 Software development process.

In the production trajectory, analysis and specification, as well as design, correspond to the interaction chain discussed above under the heading of information-technology consultancy. Code production and systems-testing merit some more explanation. Code production is the translation into actual programs of the macro-design: the "software structure" which satisfies the specifications corresponding to the information needs of the client organization. Customized software production, or programming, is divided into individual modules, these modules can be pre-existing kernels,

193

available in the software libraries of the systems developer, or they can be programmed in response to the specific needs of the clients. Customized software development tools are used to combine both system kernels and individualized programming. These tools assure coherence between different models, enforcing the norms resulting from the design stage. Each module is tested on its own, as well as in its rôle as an integral part of the overall system. In the final stage of the software development process, acceptance testing by the client involves recognition of the norms set by systems developer and client in the design stage. It requires explicit reciprocal agreement by client and supplier on quality and productivity of information flows and procedures.

The quality trajectory intervenes at all stages of the development process. It involves testing the respect of user specifications and design norms, as well as professionally accepted quality and productivity norms. Quantitative control models, integrating different stages in the development process, make a substantial contribution to coherent, process-wide quality control.

Productivity issues have become very important in customized software markets. By means of the methodological integration of different stages (specification – design – construction – testing), economies of time and quality are put into practice. However, this is only possible thanks to a highly qualified and regularly retrained staff.

As is the case for information-technology consultancy, customer involvement in systems development varies considerably among software houses and departments. Discussions on "bottom up" versus "top down" approaches are equally valid in this context, as in a consulting environment (Moulaert et al. 1990). There are serious implications for the training of end-users to maximize benefits of investment. "Bottom up" approaches, involving end-users at all levels in the client organization, require intensive organization-wide training. Insofar as end users are involved in design, programming, and resourcing of programs, then flexible user-friendly programming languages and query systems must be used. In "top down" approaches, only management (from corporate to divisional and middle management) is directly involved with analysis and design. Training efforts are much more focused. Direct co-operation between developer and client is restricted to those two stages, as well as to acceptance testing.

High-value-added network services are among the newest C&C services. Their development is directly linked to the integration of computing and communications technology, allowing for the flexible combination of data storage, processing and transmission through networks and systems of networks. The epistemology and classification of these services is not straightforward. We follow the approach of the UK DTI (1988) which, in the context of the VANGUARD program, ordered a study on the interworking of value-added and data services (VADS).[3]

The common classification of value-added and data services is based on their technical function (i.e. a horizontal approach to classification). This classification distinguishes between: communications services (e.g. electronic mail, messaging, video-conferencing), information services (e.g. teletext, databank access), and transaction services (e.g. ticket reservation, electronic data interchange, tele-shopping). This approach is too technical, however, and does not help in understanding how, through the process of information gathering, processing, transformation and communication, specific information and transaction values are created to meet the needs of specific groups of end users. This approach does not highlight the basic fact that, the more specific the information and transaction-use value of these services, the more eligible for market exchange they become.[4]

Looking at Figure 8.3, which summarizes this increasing value-added approach to network services, and following the approach developed in DTI (1988), one understands that elementary communication and messaging services do not have a "user-formatted" content. This type of communication is "only a slightly more sophisticated method of conducting normal correspondence". While the input by the users can be unstructured, the information made available by the computer is structured, quite often customized, and of strategic value: "Massive efficiencies can be achieved through electronic data retrieval, and in some instances have revolutionized businesses through the faster and greater availability of information. The value comes from the quality of information rather than from the communication method" (DTI 1988). Finally, transaction services "are bringing a fundamental change to the way in which many basic industries are running their businesses. Trading relationships are dependent upon a new way of interacting" (DTI 1988).

The possibilities for customization are many: data-management systems, mailing systems and data banks can be adapted to, or developed in relation to, user needs. Interworking between existing communication systems, private and public networks, different applications supported by different types of terminals and communication vehicles – all offer tremendous possibilities to produce the appropriate solution to client information- and communication needs. A substantially new market of "systems integration" as opened up since the mid-1980s; communications equipment producers play a much more important part in this market than in the more narrowly defined IT systems-integration market.

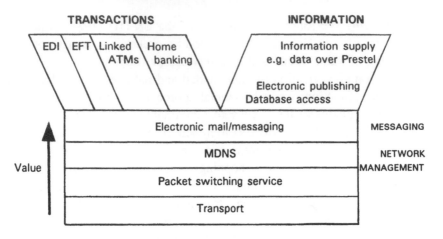

Figure 8.3 VADS classification (*Source:* DTI 1988)

The DTI study of production aspects of these three advanced C&C services stresses the huge possibilities for customized production and delivery. These possibilities pose particular problems for co-ordinating the complex of suppliers, inputs and processes, as well as the effects involved in C&C service provision. This need for project- and solution-orientated co-ordination means an additional challenge to organizational innovation. It stresses the importance of organizational dynamics as an innovative factor in supply structures and strategy.

Spatial organization of C&C service suppliers
The geography of service-supply structures and markets is complex and difficult to specify. Given the extreme diversity of suppliers, processes and products, and the limited focus of this book on services, no exhaustive inventory of manifest spatial configurations can be provided.

To begin with, market-penetration strategies in software and consultancy point to a relative decentralization of facilities. Agencies are opened, often temporarily, in response to new or growing markets; large-scale systems development and facility-management jobs lead to the location *in situ* of consultancy teams, but especially systems development teams. In a market, some segments of which show signs of saturation, high-level consultancy firms are less reluctant to explore second-rank markets (Moulaert 1988).

Secondly, C&C equipment and service suppliers are connected in many ways. Even if they are completely separate institutions, they work the same technology. Therefore, R&D co-operation is necessary with respect to

196

systems architecture, user specifications, and software development methodology and technology. This co-operation can be very explicit, in the form of joint initiatives such as project teams, often institutionalized as a joint-venture; or it can be very loose, with informal exchange of information between teams, participation in workshops, or professional publications. But the connections between equipment and service suppliers exist not only at the research level. The new philosophy of "global customized solutions" forces co-operation between professions either within or between supply organizations. Systems integration means not only the integration of products into a consistent complex, but also the proliferation of interdisciplinary co-operation in analysis and systems development.

These and other needs for close co-operation between different C&C activities and professions indicate a tendency towards economies of agglomeration and communication (Martinelli 1988). This need has been empirically confirmed for France in Moulaert et al. (1990).

Thirdly, the spectacular development of C&C systems involves the construction of specific equipment, such as data-centres, teleports and networks. It is our impression that, other than in the case of development strategies by state agencies, peripheral regions are not preferred as locations or territorial bases (in the case of networks) for such purposes. However, installation of such equipment in "second-tier" regions is not exceptional, and could become an asset for their development.

C&C technology, through its capacity to store and transmit data, allows for a certain potential mobility of suppliers and users. But the requirement for highly skilled labour and the importance of face-to-face interaction between and within R&D facilities are very important "spatial fix" factors, especially for high-level services such as consultancy, systems integration, and customized software production. It would therefore be naïve to expect C&C to overcome space by means of reducing transaction time.

C&C service suppliers:
a temporary or a structural phenomenon?

Some possibly contradictory developments have transformed the sectors of computing, communications and services over the past few years. A purely financial reading of these dynamics might lead to the conclusion that the market of the C&C integrated sector will be dominated by a limited number of C&C equipment firms, who will also have become service suppliers. However, a qualitative analysis of these developments results in a more

197

nuanced perspective.

Even if such firms as IBM and AT&T become global computing, communications and service suppliers, they have to develop technological and service assets which their original industry trajectory did not include. In practice, this means: (a) fundamental R&D, allowing for a global production and marketing strategy, exceeds the financial means of even the largest firms, and stimulates creative co-operation for specific research projects (Jowett & Rothwell 1989); (b) hardware producers and network carriers must make serious efforts to come to terms with the demand for customized services. Customization demands specialized service departments, but even more so co-operation with high-level service suppliers who have a strong competitive advantage over "manufacturers" in this respect. We think for example of leading international consultancy and software firms who, aware of their competitive strength, have not yet suffered financial absorption by the hardware giants. Even if such absorption were to develop, professional independence of service specialists must be guaranteed; if not, their skills could be lost, and with them their customized service marketing and client-focused methodology.

Ironically, the ultimate outcome of financial concentration in the C&C market might be that C&C hardware producers become predominantly service suppliers. In fact, the growth figures of services are significantly higher than those of equipment manufacture. Services already account for over 40% of turnover of some major C&C equipment producers.

Some firms in our research had adopted a strategy of service-sector expansion to compensate for declining margins on increasingly commodified hardware equipment. Nonetheless, the hardware firms must be able to convince customers that the product supplied (i.e. the ensemble of hardware/software/"orgware") is the best for their requirements. Indeed the independence of the manufacturing and service sectors is precisely the reason why ITC consultancies can bridge computers and communications or, more significantly, between hardware and services requirements.

Notes
1. Systems integration includes ITC plus the actual construction and operationalization of the IT system.
2. The major accounting problem is that very often many services, especially in engineering and systems-engineering projects, are billed without making a distinction between hardware, software and other services. Even in countries where separate accounting of services and equipment is requested, as in France, in practice the distinction is not straightforward.

3. VANGUARD is a joint UK government and industry initiative "to improve the profitability and competitive position of British industry by promoting increased use of Value-Added and Data Services (VADS)." (Department of Trade and Industry 1988)
4. This is also why this second approach is used as an analytical framework for the study of network "interworking": access to and combination of different networks, becomes interesting as of the moment the value added of the services to the potential clients is sufficiently high.

CHAPTER NINE
Global localization in computing and communications: conclusions

Philip Cooke

It is clear that important changes have occurred in the industries that have been the subject of this study. The policy arenas in which they operate in Britain and France are significantly different in the 1990s from the way they were in the 1970s, with the more obviously radical change having taken place in Britain. The technologies in computing & communications have been transformed by the widespread diffusion of digitalization, miniaturization and technological integration. These, in turn, have enabled the companies studied to develop their corporate organization. In computing & communications, more than most industries, the aspiration towards becoming dynamically flexible in research, innovation, production and marketing has been most closely approached. Computer services firms, centrally engaged in their most developed form with engineering new corporate management structures are – as well as displaying substantial dynamic flexibility themselves – diffusing the capacity to enhance it in their client organizations.

These new requirements have important implications for corporate workforces. It is clear that the 1980s saw a notable shift in the occupational structure of the computing and communications industries. Put at its most general, this represented a shift towards a more skilled technical, professional and scientific occupational profile and away from a more typically Fordist employment structure with, at the base, a large category of less-skilled, routine, assembly workers. Accompanying this marked occupational shift has been something of a spatial shift in industrial location. From having been distinctly metropolitan industries, computing & communications have developed a skewed, bipolar spatial distribution. In France we see the emergence of a southern pole; in Britain a less focused, non-metropolitan development tendency. In both the British and the French cases, part of this trend can be explained by the locational behaviour of inward investors. Computer services, by contrast, remain strongly metropolitan in their locational characteristics, with even a suggestion that in Britain this is becoming a more pronounced feature of the industry. It is, in any case, an industry with high demands upon skilled technical and professional labour, a resource that tends to be concentrated in metropolitan areas.

What shapes the market: governments, technologies, producers or users?

In introducing this book, three main aims were stated in undertaking the research that it reports. The first of these was to explore the relative weight of government policies versus sectoral dynamics in explaining the developmental profiles of the computing & communications industries in Britain and France. The counter-position of these two forces is of course artificial, since the two are, to varying degrees, intertwined. Nevertheless, the existence of two very different policy regimes in the two countries made it worth comparing industry performance in this light.

French computing & communications industries remain subject to very close involvement by the state: the telecommunications service is publicly owned and, while on the face of it there has been significant movement in the equipment supply industry, two main suppliers with strong links to an earlier generation of equipment suppliers remain in place. In computing, France's main company, Bull, is seeking to establish a global position, but depends heavily on state funding to do so. Britain's telecommunications service has been privatized, competition of a limited kind has been introduced, as has regulation of the market. The equipment suppliers have been subjected to severe competition and have in one sense disappeared as independent entities. Each of the three established suppliers – GEC, Plessey and STC – have become part of foreign-owned corporations, with only GEC retaining some degree of identity as the albeit smaller partner in the merger with Siemens. The same can be said of Britain's main computer hardware producer, ICL, now a subsidiary of Fujitsu. Whereas French policy has

201

been to seek to build French world champions, with varying degrees of success, the effect of British policy has been to let the market dictate the nature and extent of British participation in the global strategies of world-class firms from outside Britain.

On the face of it then, it appears that state policies have been crucial to the different trajectories taken by C&C equipment firms in the two countries in question. While the French have nurtured their national champions and nursed them to international prominence, the British appear to have cold-shouldered theirs and left them to survive in a harshly competitive environment as best they can. To focus on the French case first, there has been a state industry policy which, despite governmental changes, has given a high degree of security to firms in dealing with the turbulent world in which C&C firms operate. From the 1970s when the French government showed it was taking the promised "informatics revolution" seriously, there has been a degree of consistency towards the industry that firms elsewhere might well have envied.

The modernization of DGT, turning the French telecommunications system into one of the most advanced in the world, compared favourably with the troubled history of British Telecom prior to its separation from Britain's PTT, the General Post Office, and its subsequent privatization. Investment levels by DGT were almost twice those of its British equivalent over the 1975–85 period; it became a more commercially sophisticated business, competition between suppliers was enhanced, and R&D became a leading priority. Thereafter, the equipment-supply industry was restructured, with the French firms Alcatel and Thomson replacing Ericsson and ITT as the dominant suppliers. Subsequently, France Telecom (as DGT was renamed) developed new network services early, and liberalized the supply of equipment to customer premises. Thereafter, the main suppliers were nationalized and in 1983 Alcatel was made the main supplier, following an agreement to transfer Thomson's telecommunications competences to Alcatel. With the acquisition of ITT Europe, Alcatel became a global corporation, second in the world only to AT&T as a telecommunications equipment firm.

Interestingly though, and despite these apparent transformations, it is worth bearing in mind that, although Alcatel is now a global corporation, as a result of its acquisitions it embodies one of the two old suppliers to DGT, namely ITT. Moreover, it is clear that, within Alcatel, the former German subsidiary of ITT, Standard Elektrik Lorenz (SEL), is seen as something of a jewel in the corporate crown. So, in a sense, in a new enlarged European form ITT remains, in its afterglow, the key French supplier. Moreover, Ericsson has re-emerged as the second, smaller supplier to France Telecom through its alliance with Matra, which received the switching activities of CGCT in 1987 and has become a rapidly growing equipment supplier to the French market. The differences are ultimately those of ownership and scale: French (despite Alcatel NV's temporary decampment of its headquarters abroad) and global in extent. French government policy has been of central importance to these developments, but the developments themselves have been occasioned by the practices of competitors in the industry who are driven by costs and the insatiable demand for global market share to recoup them.

The case of the British telecommunications equipment-supply industry reveals the imperatives of globalization equally vividly, but from the perspective of the acquired rather than the acquirer, and in the absence of state support of a direct kind for the creation of global champions out of former national champions. Technological difficulties with System X (the new digital switching technology) were a symptom of the relative failure of the established, cosy, relationships of the Keynesian mixed economy in telecommunications, with a monopoly carrier and a "club" of privileged suppliers. The "club" was powerful and politically influential, able to resist attempts to improve efficiency which might entail competition and rationalization. By 1984, with the privatization of BT, this system had been effectively dismantled. Ericsson was allowed to sell public switches in Britain, causing an immediate loss of market share to the System X suppliers, GEC and Plessey. Ericsson took 30% of the domestic digital exchange market.

This, in turn, entailed a significant restructuring of the supply industry. STC had been excluded from System X; Plessey had become the main contractor, with GEC as its prime subcontractor. This eventually led to the two System X suppliers forming a joint venture, following GEC's first abortive bid for Plessey. Scale was enhanced, but GPT (as the joint venture was called) remained small in global terms. Moreover, it had only a miniscule share of the market outside Britain. Ultimately, the global imperative has resulted in GPT becoming part of the Siemens telecommunications equipment supply empire as a by-product of the joint GEC–Siemens bid for Plessey which, in 1989, was successful. STC meanwhile, forced into a niche-market strategy for its switching activities, has been acquired by the burgeoning Canadian switch-maker, Northern Telecom.

What can be concluded from these divergent British and French experiences? Three of the most obvious points are the following.

First, successive French governments have sought to place a national imprint on processes in telecommunications which they were early to anticipate. While it remains to be seen whether Alcatel will be able to continue as the major European corporation, or a significant global corporation, in its own right (given the power of Siemens), let alone Japanese and American competitors, French state policies have clearly demonstrated the medium-term advantages of good planning. The British position has been to eschew planning and to stimulate the market to effect the restructuring of the supply industry. In the process, British firms have become more globalized by acquisition. In the longer term this runs the risk of British firms being "hollowed-out" by becoming mere assemblers and distributors of foreign technology. On the more positive side, the attachment of British equipment suppliers to Alcatel's European rival Siemens and the expanding Canadian firm Northern Telecom means that they are at least involved with firms possessing excellent technologies and global strategic intent. Alcatel has recently pulled out of the US market, while Siemens retains a presence there. Moreover, Siemens is a richer, larger and more deeply integrated company in C&C technologies generally. The signs, therefore, are that GEC–Siemens will be the leading European global competitor in telecommunications equipment.

Secondly, despite the policy environment within which British and French equipment suppliers operate, technological change has forced the key companies into strategies of corporate concentration. The race to develop the first digital public switch, and thereby tap the enormous global market for this advanced technology, resulted in British firms being poorly placed by comparison with US, French and German companies. However, the costs of development of not only the first, System X, let alone the second, System Y, generation of switches have meant that even the largest corporations have both been pulled out into seeking a presence in the Triad markets – with Japan the most difficult to penetrate – and into forming alliances, not so much to develop the technology as to gain access to the distribution networks of the markets being targeted. Moreover, technological convergence of C&C technologies has forced firms to seek to work with others in the complementary technology in order to meet market demand. Neither Britain nor France is as well placed as Germany (in the form of Siemens) to achieve this convergence. Siemens, as well as being a computing and components company, has a strong and multi-faceted alliance with IBM to further the latter's telecommunications equipment ambitions and improve Siemens's weaker computing capacity.

Thirdly, the relationship between producers and users of tele-

204

communications equipment has changed. In the past, producer-power was overwhelming as national champions dictated the pace of technological change and the availability of products. Latterly, though, there has been a growth in user-power. This is partly because of the existence of a newly competitive environment following deregulation and privatization in some countries. It is also partly because of the availability of newer, more flexible telecommunications technologies and services. But, perhaps more than anything, the markets for new technologies and services have grown enormously. This has been particularly the case in the globalized financial services industries, but demand has also grown in manufacturing industries, as they integrate their international management and supply systems such that even relatively small firms involved in the manufacturing supply chain may expect to have to be electronically integrated with their customers. Demand for customized solutions forces equipment-supply firms to be adaptable.

In brief, markets are presently triumphing over government policies, with market edge being appropriated by firms with the leading technologies. This judgement is even more appropriate in the case of the computing industry. In Britain, the era of government assistance to the industry, particularly ICL, has long passed. ICL is now owned by Fujitsu, and it is instructive that, as a consequence, it has now been forced to withdraw from the European Round-table of Information Technology Producers. This may be an indication of the likely posture taken by the European Commission towards ICL's status as a bidder for science and technology research funds under programmes such as ESPRIT. Moreover, now that it is Japanese-owned, the question of ICL's role as a supplier of military technology to the Ministry of Defence has yet to be resolved. In France, the development of Bull echoes that of Alcatel, to some extent: a process of attempted globalization by acquisition. The acquisition of the computing activities of the US firm Honeywell signified the earnestness of that ambition. But Bull's holding company is state-dependent to a remarkable degree, a condition that must change in the light of European Commission competition policy. As a consequence, whether Bull can survive as Europe's major indigenous computing company remains open to question.

Technological change has made the computer hardware business extremely competitive, especially as miniaturization has made the mainframe power of yesteryear available on desktop equipment. User-power is exceptionally strong as hardware had become something of a commodity, leaving equipment producers increasingly dependent on their systems-integration and service-supply activities to retain competitive edge.

In the face of massive competitive strength from Japan and the USA, the future of a European computing presence probably lies in a combination of alliance strategies and the development of special competences in niche markets. Perhaps, once again, the British approach better reflects these realities than does the French.

Finally, the burgeoning computing & communications services industry is a highly market-driven phenomenon. Three features of the industry are worth noting. First, as already suggested, computer hardware producers are reaping significant advantage from their capacity to configure systems to meet specific customer-need. This will doubtless continue. Secondly, the vast range of specific applications means that there will continue to be an important rôle for niche-suppliers. However, many of these are under-capitalized and labour-intensive, hence concentration is on the agenda for many, if it has not yet already taken place. Thirdly, where C&C services activities have been absorbed into large management consultancies or where mergers have produced significant scale amongst software and systems-engineering houses, such firms are having an important impact upon corporate reorganization by client firms.

The pervasiveness of information technology means, as is shown in Chapter 8, that, in order to capture value from IT investment, corporations are having their internal organizational functions substantially restructured around core information-technology competences. This then magnifies an enormous skills gap within firms, which itself adds to the market for consultants who may temporarily be brought in to help fill it. Overall, therefore, it is fairly clear that the dynamics of competition and market expansion predominate over the effects of government policies in the hardware and software segments of the C&C industries. The spectrum now ranges from the liberalizing telecommunications industries to the highly commercial computer services industries. And while the former pole of the spectrum has been significantly stimulated by deregulatory tendencies in government policies – even, to some limited extent in France – the latter pole signifies the "vanishing point" in the developmental profile towards which a large part of C&C industry is, in any case, headed.

The regulationist perspective and research on C&C industries

Providing a theoretical backcloth to the research reported in this book has been the work of the "regulation school". One of the aims of the book has been to evaluate this conceptual framework for purposes of doing empirical research. This is, in some ways, an apparently straightforward task, but on further inspection it becomes more difficult. The principal difficulty is that, while the perspective adopts the language of regulation (and by inversion, also deregulation), the ways in which the concept "mode of regulation" is used are at once highly general and macro-theoretical as well as, to say the least, rather loose. By contrast, the regulation and deregulation that characterizes the experience of some telecommunications service suppliers and, by extension, some of their equipment suppliers, can be quite precisely defined (even though deregulation itself is also often bandied about loosely as a generic term). In its essentials, regulation is a set of governmentally instituted rules which govern the practices of firms to ensure that, where competition is absent, nevertheless monopoly powers do not infringe the public service interest; or, where competition does exist, market principles should not override certain norms of public service.

Resolving the general and the specific meanings of the concept "regulation" is thus not easy. Nevertheless, the narrower definition is compatible with the broader inasmuch as the former is clearly one of the many categories of the latter. Another category, for example, would be the manner in which in Britain in the 1980s and early 1990s the economy has been regulated to a large extent by governmental variation of interest rates. The problem arises particularly where the industries that are the focus of empirical research have not, in any abiding way, been "regulated" by government – the computer industry being a case in point – even though they may at various points have been recipients of life-saving, or life-enhancing, government subsidy, as with ICL and Bull. This is even more the case for computer services industries, where, except for those firms which rely heavily on defence contracts, regulation in any meaningful sense of the word has never been a significant factor in their existence.

In cases like these the value of the regulationist perspective is circumscribed by the relative absence of direct state involvement, though the concept of "the mode of regulation" is not confined solely to state activity. And it is perhaps here that ideas associated with the regulationist approach help more than most such macro-perspectives to link developments in the institutional framework, within which unregulated firms operate, back to developments in the accumulation regime. For it is in the institutional framework that the vaguest of all regulationist insights proves to be quite subtle and penetrative. The "consumption norm" is the concept in question. There has unquestionably been a re-evaluation of the consumption norms of quality, reliability and adaptability during the period to which some regulationists refer as post-Fordist. Information technology, of which computing & communications are the network infrastructure, can improve these in their own right. But, for the consumers of IT, the real value added from the investment comes from the reorganization of the firm such that it can maximize the advantages offered by IT. The "Solow paradox" (see discussion in OECD 1988c), which identifies the impact of IT everywhere except in corporate productivity figures, is resolved when it is recognized that computing and communications improvements in themselves may be realized only when the more intangible resources of organizational, educational and skills reforms have been put in place in ways which enable the full potential of IT to be attained.

The rise of user-power is a strong element causing equipment and services suppliers (but particularly the former) to become more "market-facing", the term many of the British-based respondents used in explaining the ways in which their corporate reorganization had been slanted in the late 1980s. This meant significant changes being instituted in both corporate strategy and corporate organization. Corporate strategy shifts included improving the level of detailed information available from customers, expending more resources on providing aftercare to customers, being better informed about the prospects for expansion into new, non-domestic markets and new kinds of markets (e.g. services as well as hardware), and being prepared to engage in what Ericsson referred to as "value-added partnering" with competitor companies as well as complementary companies overseas and domestically. Finally, in most cases, corporate strategy, had already involved shedding acquired non-core competences and "getting back to basics", i.e. the key product–service focus of the company. Around core competences the innovation effort was evidently being upgraded by both increases in innovation expenditure in-house and closer interaction with innovative suppliers.

Corporate organization in the supplier companies had also generally been subject to reform. For example, the commitment to be more "market-facing" had often meant redeployment of key personnel, often in substantial numbers, from back-office to front-office marketing activities. The need to be more competitive in a world market, where sometimes nationally protected firms had been exposed to the harsher climate of foreign competition, had meant reductions in in-house production and design with, in virtually all cases, an increase in use of the external supply-chain of subcontractors. While the number of first-tier subcontractors may in most cases have declined, the move to closer "preferred" status for key suppliers had resulted in significant increases in the value of materials now bought in, as compared with the position five years before. In some cases – most obviously ICL – experimentation with not only computer-aided manufacturing but the much more difficult computer-integrated manufacturing were well advanced and, in ICL's case, installed. This approach depended fundamentally upon the availability of the appropriate C&C technology which, as well as integrating the internal organization of production, integrated first-tier suppliers electronically into the production process as well by means of electronic data interchange systems.

But perhaps most interestingly of all, in the light of some of the vaguer theoretical notions regarding post-Fordist flexibility as discussed in the regulationist literature, there appear to be strong signs of the emergence of forms of "dynamic flexibility", whereby the permanent innovation process pioneered by the Japanese was being aspired to, and in some cases approached. The concept of "knowledge-intensive production" was widely understood in the target firms, and many were well advanced in the development of overlapping product life cycles to the extent that, for some products, product life cycles were approaching a parallel position. That is, the next generation of product was getting close to being put into production as the previous one was being introduced.

For this to be happening – and the degree to which it *was* varied by firm, product and market-dynamism – the level of intra-corporate integration had had to be improved. Thus, for example, in response to the question of whether firms had adopted matrix-management techniques – lateral project-based lines as well as vertical command-based lines – most either had, or had gone beyond them to a more strongly project-focused management structure, or had developed some hybrid with matrix-like elements to it. The other organizational innovation which had developed was the greater use of networking; either customer–supplier networking with higher degrees of trust and supplier project-involvement, or competitor-networking through various forms of alliance, joint venturing or partnership arrangement.

Now, each of these corporate organizational or strategic reforms is consistent with regulationist insight: that the Fordist model of hierarchical, standardized mass production has weakened and is being transcended by a more decentralized (managerially), less standardized (more systems-integrated) form of mass production which, as a delivered system, may be highly customized to user requirements. A new technological paradigm has taken shape around both C&C industries and their insertion into production and management systems in other sectors. The relationship of firms to the market has changed in that user-power has grown and the consumption norm has become more diverse and demanding. In the process the Fordist worker is, if not in retreat, then certainly in numerical decline and, as the supply chain extends and diversifies, is found in a more vulnerable labour-market setting. Competition is fiercer but, to some extent, more subject to mitigation by corporate partnering practices. Financial resources, required in greater amounts in real terms than ever before, are available, in part due to the relative relaxation of credit controls under contemporary monetary policies. Governments have, to varying degrees, liberalized industries they once regulated most closely via public control. And, whether through liberalization of aspects of supply to a publicly-owned PTT, as in France, or privatization as in Britain, the trend has been towards more globalized markets and production structures. Added together, these constitute a sea-change in the production regimes of the computing & communications industries.

However, the regulationist perspective is open to question in the sense that it integrates the accumulation regime with the supportive or guiding regulatory mode. Now, as we have seen, there is a strong sense in which we have argued for such interaction, but with corporate dynamics uppermost. But with two such divergent regulatory regimes as the French and British, it is hard to see how they can both be equally supportive. That is, if both neo-liberalism and *dirigisme* are equally supportive, does the regulatory regime matter at all in the long run? Maybe not; perhaps regulatory regimes derive from systemic political perceptions of national interest, refracted through a lens which reveals only dimly the complex dynamism of technological and industrial change. It should be remembered that the "regulation school" (of which there are at least three in France alone) is a specifically French group of theorists, looking at France and the rest of the world from a viewpoint in which, perhaps uniquely in the West, the state is immersed in corporate affairs, and without which French capital may not have been able to withstand pressures of foreign competition. This is not to comment on the efficiency or effectiveness of French industry, but rather its scale and resources in relation to those of Germany, Japan and the USA. Without *dirigisme*, French C&C capital would almost certainly have experienced the same fate as British, i.e. absorption into the major global corporations in one form or another. So, to that extent, and in the short term, state-regulation in the broad sense has made a difference.

Finally, what seems to count for a major portion of the sea-change in the organization of production in C&C industries in both countries is at least as much user-power as producer-power. The regulationist perspective is one which, in the final analysis, privileges the production sphere (technological paradigm plus labour process) over anything else in explaining major changes in capitalism. Whether or not C&C industries are anomalous is unclear; they are certainly advanced. But it is clear that, without markets through which to recoup investment in technological change and labour reorganization, those changes may be made only in vain or not at all. The key functionaries in the system of production, at least in the C&C industries, need to listen intently to the information transmitted from the market to know which technological and labour process adaptations to make in order to survive.

Global localization: presence or prospect?

Having assessed the value of the regulationist approach to the research in question, and having, on balance, found its insights regarding the nature of change in the accumulation regime more persuasive than its proposals about the synchronous relationship of regulatory modes to such regimes, it is nevertheless worth recalling that the approach is ahead of the field in its spatial analysis. That is, despite the uneven quality of the concept of a mode of regulation – fitting well in some countries for some institutional activities, less well in others for alternative institutional arrangements – it does offer an important connection from the aspatial universe of the economist to the spatial one of the economic geographer, regional scientist or urban & regional studies specialist. Put crudely, it was the regulationist perspective – barely yet formulated as such – that enabled theoretical and subsequent descriptive innovation to occur in which a coherent explanation for the Fordist spatial division of labour could be provided. Put equally crudely, the conceptual apparatus of the regulation school points unerringly to certain tendencies in contemporary corporate restructuring which may have different spatial implications.

Of the first importance here is the idea of the expansion of capital with an associated deepening and intensification of the relationships of the corporate sector, both internally and externally. By this is meant the idea of a sequence of intensities of development which characterize the deepening and developing grip of capital upon particular geographical spaces. In its extensive phase, Fordism operated as an advanced form of production, capable of expanding the market by the devices of its leading exponents and their emulators. In its intensive phase, far more external institutional and regulatory support activity is brought into play, as is

211

detailed financial control, foreign investment and the deepening of Fordist structures of production. This is an international system of production based on national economies, some of which became export bases for further market expansion on an international scale. Now, it is possible to think of post-Fordism as having been in an extensive phase since the peak of the Fordist boom in the late 1960s up to the 1990s. That is, we have seen a vast expansion of multinational capital from particular national territories, establishing production platforms overseas but keeping most headquarter functions on shore. Neo-Fordist techniques of human relations management of the labour process, the introduction of electronically automated equipment, and so on, signify the onset of problems with classic Fordism, but also the newer phase of production organization straining to be born. The 1970s and early 1980s are a period of extended crisis and much experimentation which actually results in the recognition of the superiority – in the automobile industry at least – of what Womack et al. (1990) call "lean production". Lean production is characterized by similar bases in the *integration* and *networking*, with a supply-chain focus, that we have begun to identify in the computing & communications industries.

One of the key features of our larger target firms was their common interest in and pursuit of a globalization strategy. Now, it is the combination of a *globalization* imperative with the integration imperative that gives rise to the enhanced networking propensity. Knowledge-intensive production and permanent innovation, as goals for Western companies and achievements by leading Japanese companies, could be serviced from the domestic base, were it not for one key element of the findings of this book: the heightened importance of markets. In particular, the differences between the so-called Triad markets are what give rise to the pressure for leading-edge firms to move towards "global localization". This may become the intensive phase of post-Fordist accumulation. As competition intensifies, the nuances of market culture make it more and more difficult to satisfy the more discerning user without having the capacity to tailor a generic product to the finest degree possible to meet local demand. This is least a problem in the USA, though by no means an insignificant one, and most a problem in Japan. From the power base of a market loyal to home-products, the Japanese firms Sony, Honda, Fujitsu and Mitsubishi, amongst others, have the strategic intent, as Chapter 4 showed, of achieving "local assimilation" within a continental market. Obviously, to the extent this can be achieved, it will require the intensive integration of the whole value-chain of production from R&D to marketing, not forgetting the supply-chain dimension, to be continentally complete. Global

212

localization, then, may well be the emergent intensive form of the post-Fordist regime of accumulation.

However, moving to the other pole of the geographical scale, the subnational, meso-level of the region, it is less than clear that, at least in the C&C industries, the localization dimension of the global–local couplet is present. This is not all that surprising. First, the Fordist spatial division of labour is only just beginning to show signs of erosion, with the decline of the Fordist "mass-collective worker" or operator grade. However, that decline has been precipitous, particularly in Britain, as has the massive relative increase in the knowledge-intensive occupations in both countries. The push out of the metropolitan centres to their surrounding region, and a few preferred outer regions, is also, in part, a sign of the quest for knowledge-intensive labour; both Scotland and Rhône-Alpes have an advantage in this respect.

Nevertheless, while our research regularly turned up evidence that firms were increasingly reliant upon the external supply chain, there was little evidence that one of the key expectations of some post-Fordist spatial theorists was being fulfilled, namely that this would take the form of local clustering in Marshallian "industrial district" form. However, that expectation is too severe in mature industrial economies. What seems to be a more realistic expectation is this. In France the Paris region acts as one enormous "industrial district" for C&C, as for most other French industries. However, Rhône-Alpes, in particular, seems to be developing some of the same characteristics on a much smaller scale. As Dunford (1989) and Moulaert et al. (1988) have shown there is substantial and increasing interaction between localized customers and suppliers for computer services in particular, but also computer componentry. The position in Britain is that, to the extent there is localized supply, it is throughout southern Britain, the key triangle being drawn by the M1, M4 and M5 motorways. Many respondents noted the "conveyer-belt" effect of these links and the important location factor of appropriately skilled labour at its south-eastern corner.

But despite this, and consistent with the globalization thesis, it must be noted that a larger proportion (by value) of component supply in C&C hardware industries is sourced globally. Even firms that design and produce their own semiconductor devices are forced to buy from outside their country of origin if, as is increasingly the case for even application-specific devices, they can be bought more cheaply from Japan, if not California. Thus, for the moment – and to the extent that post-Fordism, as it has been described, rooted in integration, networking and dynamic flexibility, is the

213

lean successor of Fordism – it is incontrovertibly in its extensive phase. A few leading-edge non-Fordist corporations, mainly Japanese, are professing a global-localization strategy, but scarcely any are really there yet.

What, therefore of Britain and France, at least in the advanced technological fields of computing & communications? British firms are leaner, but can be said to be global only by virtue of having been acquired (and then not necessarily for their technological excellence, except in a few cases) by firms with a greater claim to be becoming global; Fujitsu and Siemens are cases in point. British C&C firms will probably become first-tier suppliers to these global aspirants, and that, in fact, may well be the not wholly unacceptable fate of British manufacturing in general and electronics in particular. The French have taken a bold, long-term gamble that by judicious state planning they could secure a global future for one world-class firm in each of the computing & communications industries. Thus far, significant progress has been made, but Alcatel, which is the stronger of the two national champions, has failed to break out of Europe into the other Triad markets. Indeed, it has retreated from the USA, unlike Siemens. Bull, the computing national champion is in a less comfortable position, having acquired one of the USA's ailing giants, Honeywell, and it is dependent on state funding. This is unlikely to continue in post-1992 Europe. France has protected its national champions, which may be what it most wanted to do in the final analysis. Britain has effectively lost hers, whether by design or by accident it is difficult to tell.

REFERENCES

Abernathy, W. J., K. B. Clark, A. M. Kantrow 1983. *Industrial renaissance*. New York: Basic Books.

Abernathy, W. J. & J. M. Utterback 1978. Patterns of industrial innovation. *Technology Review* **80** (7) (January–July).

Adler, P. 1985. Managing flexibility: a selective review of the challenges of managing the new production technologies potential for flexibility. (July) Stanford University mimeo.

Agence de l'Informatique 1986. *L'état d'informatisation de la France*. Paris: Economica.

Aglietta, M. 1976. *Régulation et crises du capitalisme: l'expérience des Etats Unis*. Paris: Calmann-Lévy.

Amin, A. & J. B. Goddard 1986. *Technological change, industrial restructuring and regional development*. London: Allen & Unwin.

Arnold, E. & K. Guy 1986. *Parallel convergence: national strategies in information technology*. London: Pinter.

Aurelle, B. 1986. *Les télécommunications*. Paris: La Decouverte.

Bar, F. (forthcoming). Telecommunications networks: exploring new infrastructure concepts. In *Regulation, innovation and spatial development*, P. Cooke (ed.). London: Sage.

Barreau, J. & A. Mouline 1987. *L'industrie éléctronique française: 29 ans de rélations État-groupes industriels, 1958–1986*. Paris: Librairie Generale de Droit et de Jurisprudence.

Beckouche, P. 1987. L'industrie electronique française: les regions face à la transnationalisation des firmes, PhD dissertation, Université de Paris I Pantheon Sorbonne. Paris: UER de Geographie.

Bertho, C. 1981. *Télégraphes et téléphones: de Valmy au microprocesseur*. Paris: Livre de Poche.

Boyer, R. 1986. *La théorie de la regulation: une analyse critique*. Paris: La Decouverte.

Boyer, R. 1988. Technical change and the theory of regulation. In *Technical change and economic theory*, D. Gosi et al. (eds), 67–94. London: Pinter.

Boyer, R. & B. Coriat 1986. Technical flexibility and macrostabilisation. Paper at the Venice Conference on "Innovation Diffusion". Venice, 2–4 April.

Boyer, R. & J. Mistral 1983. *La crise actuelle: d'une analyse historique à une vue prospective* (8304). Paris: CEPREMAP.

Caulkin, S. 1987. ICL's Lazarus act. *Management Today* (January) 56-63.

Chandler, A. 1966. *Strategy and structure: chapters in the history of American industrial enterprise*. New York: Doubleday.

Chandler, A. 1977. *The visible hand: the managerial revolution in American business*. Cambridge, Mass.: Harvard University Press.

Chesnais, F. 1987. Les accords de cooperation technologique et les choix des entreprises européennes. Colloque CPE-FAST-CGP. 23-25 April.

Clarke, L. 1986. Making change in ICL. *Management Today* (May) 76-80.

Cohen, S. & J. Zysman 1987. *Manufacturing matters*. New York: Basic Books.

Cohendet, P. & P. Llerena 1989. Flexibilité, risque et incertitude dans la théorie de la firme: un survey. In *Flexibilité, information et décision*, P. Cohendet & P. Llerena (eds). Paris: Economica.

Communications Week International 1989. Forging new links. *Communications Week International* editorial, 11 December.

Cooke, P. 1988. Flexible integration, scope economies and strategic alliances: social and spatial mediations. *Society and Space* 6, 281-300.

Cooke, P. (ed.) (in press). *Regulation, innovation and spatial development*. London: Sage.

Cooke, P. & K. Morgan 1990. *Learning through networking: regional innovation and the lessons of Baden-Württemberg*. Regional Industrial Research Paper 5. Cardiff: University of Wales.

Coriat, B. 1988. Flexibilité technique et production de masse. De la socialisation flexible à la flexibilité dynamique. Colloque international sur la theorie de la regulation. Barcelona. June.

Crandall, R. W. & K. Flamm 1989. *Changing the rules: technological change, international competition and regulation in communications*. Washington DC: Brookings Institution.

Daniels, P. 1988. Les services aux entreprises et le développement de l'économie spatiale. In *La production des services et sa geographie*, F. Moulaert (ed.). Special issue of *Cahiers Lillois d'Économie et de Sociologie*.

Daniels, P. & F. Moulaert (eds) 1990. *Producer services: an agglomerated division of labor*. London: Pinter (Belhaven).

Daniels, P. W., N. J. Thrift, A. Leyshon 1989. Internationalisation of professional producer services: accountancy conglomerates. In *Multinational service firms*, P. Enderwick, (ed.), 79-106. London: Routledge.

De Jong, H. W. 1989. Mergers and takeovers: theory, history and empirical evidence. 4th session, Mediterranean Summer School on Industrial Organization. Caroese. September.

De Olivera, G. 1981. L'influence du progrès technologique sur l'emploi dans l'industrie des télécommunication. In *Les mutations technologiques*, ADEFI, 30-48. Paris: Economica.

Delapierre, M. & C.-A. Michalet 1989. Vers un changement des structures des multinationales: le principe d'internalisation en question. *Revue d'Économie*

Industrielle **47**.

Delapierre, M. & L. K. Mytelka 1988. Décomposition, recomposition des oligopoles. *Économie et Société* **11–12** (November–December) 31.

Dicken, P. 1986. *Global shift: industrial change in a turbulent world*. London: Harper & Row.

Doran, P. 1986. How to achieve performance. *Management Today* (April) 94–100.

Dosi, G. 1982. Technological paradigms and technological trajectories. *Research Policy* **14**.

Dosi, G. 1988. The nature of the innovative process. In *Technical change and economic theory*, G. Dosi et al. (eds), 221–38. London: Pinter.

DTI [Department of Trade and Industry] 1988. *Value-added and data services interworking study*. London: HMSO.

Dunford, M. 1989. Technopoles, politics and markets. In *Strategies for new technology*, M. Sharp & P. Holmes (eds), 130–52. Hemel Hempstead: Philip Allan.

Dunning, J. & J. Cantwell 1987. *IRM directory of statistics of international investment and production*. London: Macmillan.

Durand, P. 1974. *Industries et régions*. Paris: La Documentation Française.

Dussauge, P. & B. Ramanantsoe 1987. *Technologie et strategie d'entreprise*. Paris: McGraw-Hill.

Encaoua, D. & P. Koebel 1987. Règlementation et dérèglementation des télécommunications: leçons anglo-saxonnes et perspectives d'évolution en France. *Revue Économique* (2) (March).

Enderwick, P. 1989. Some economies of service–sector multinational enterprises. In *Multinational service firms*, P. Enderwick (ed.), 3–34. London: Routledge.

EITB [Engineering Industry Training Board] 1989. *Industry profile: British engineering – employment, training and education*. Stockport, Cheshire: EITB.

ERMES 1988. La demande de services complexes des firmes multinationales et l'offre correspondante. Lille: Université de Lille I, Faculté des Sciences Économiques et Sociales.

Ernst, D. 1987. *Global competition, strategic alliances and the worldwide restructuring of the electronics industry* (mimeo). Paris: OECD.

Flamm, K. 1987. *Targeting the computer. Government support and international competition*. Washington DC: Brookings Institution.

Flamm, K. 1988. *Creating the computer: government, industry and high technology*. Washington DC: Brookings Institution.

Freeman, C., J. Clark, L. Soete 1982. Unemployment and technical innovation. London: Pinter.

Freeman, C. & C. Perez 1988. Structural crises of adjustment: business cycles and investment behavior. In *Technical change and economic theory*, G. Dosi et al. (eds), 38–66. London: Pinter.

Frobel, F., H. Hinrichs, D. Kreye 1980. *The new international division of labour*. Cambridge: Cambridge University Press.

Gadrey, J. 1989. Nouvelles strategies de l'offre de conseil aux entreprises. *Revue Française de Gestion*, November–December.

Gadrey, J. 1990. Les systèmes d'emplois tertiaires au coeur des transformations du marche de travail. *Formation Emploi*, 1st trimester 1990.

GESI 1986. *Grappes technologiques: les nouvelles strategies d'entreprises*. Paris: McGraw-Hill.

Gilles, B. 1978. *Histoires des techniques*. Paris: Gallimard.

Gordon, R. & L. M. Kimball 1987. Beyond industrial maturity: planning the future of Silicon Valley. GREMI Conference on Innovation Policies at the Local Level, Paris, 14–15 December.

Granovetter, M. 1985. Economic action and social structure: the problem of embeddedness. *American Journal of Sociology* **91**, 481–510.

Grey, S. J. & M. C. McDermott 1987. International mergers and takeovers: a review of trends and recent development. *European Management Journal* **6** (1), 26–43.

Grubel, H. G. 1989. Multinational banking. In *Multinational service firms*, P. Enderwick (ed.), 61–78. London: Routledge.

Guile, R. & H. Brooks 1986. *Technology and global industry*. Washington DC: National Academy Press.

Hacklisch, C. 1986. *Technical alliances in the semiconductor industry*. New York: New York University, Center for Science & Technology Policy.

Hagedoorn, J. 1989. Organizational modes of inter-firm co-operation and technology transfer (MERIT paper). Maastricht: State University of Limburg.

Hagedoorn, J. & J. Schakenraad 1988. Strategic partnering and technological cooperation (MERIT paper). Maastricht: State University of Limburg.

Hakansson, H. 1989. *Corporate technological behaviour: co-operation and networks*. London: Routledge.

Hamel, G., Y. Doz, C. Prahalad 1989. Collaborate with your competitors and win. *Harvard Business Review* **69**, 133–39.

Hamill, J. 1988. British acquisitions in the United States. *National Westminster Bank Quarterly Review* (August) 2–17.

Harrigan, K. 1985a. *Strategies for joint ventures*. Lexington, Mass.: Lexington Books.

Harrigan, K. 1985b. Vertical integration and corporate strategy. *Academy of Management Journal* **28** (2).

Hart, H. 1945. Logistic social trends. *American Journal of Sociology* **50**, 337–52.

Hills, J. 1984. *Information technology and industrial policy*. London: Croom Helm.

References

Janssens, F. 1985. Van chronometer tot robot industriele arbeid en bedriffsorganisatie in beweging. In *Het laboratorium van de crisis*, E. Polekas (ed.). Leuven: Kritak.

Jowett P. & M. Rothwell 1989. *The economics of information technology*. London: Macmillan.

Kelly, T. 1987. *The British computer industry: crisis and development*. London: Croom Helm.

Klein, B. H. 1984. *Prices, wages and business cycles*. New York: Pergamon Press.

Klein, B. H. 1986. Dynamic competition and productivity advance. In *The positive sum strategy*, R. Landau & N. Rosenberg (eds), 112–28. Washington DC: National Academy Press.

Kline, S. J. & N. Rosenberg 1986. An overview of innovation. In *The positive sum strategy*, R. Landau & N. Rosenberg (eds), 10–32. Washington DC: National Academy Press.

Knobel, L. 1988. Hewlett-Packard's culture shock. *Management Today* (June) 100–1.

Kristensen, P. H. 1986. Industrial models in the melting pot of history. (August) Roskilde University mimeo.

Landau, R. & N. Rosenberg 1986. *The positive sum strategy*. Washington DC: National Academy Press.

Leborgne, D. & A. Lipietz 1988. New technologies, new modes of regulation: some spatial implications. *Society & Space* 6, 263–80.

Lemattre, M. (nd[a]). *Emploi et qualification dans les industries informationnelles*. UFR des Sciences Économiques et Sociologiques, Université de Liile Flandres-Artois mimeo.

Lemattre, M. (nd[b]). *Repartition régionale des industries informationnelles* (mimeo). UFR des Sciences Économiques et Sociologiques, Université de Lille Flandres-Artois.

Lipietz, A. 1984. *Accumulation, crises et sorties de crise: quelques réflexions méthodologiques autour de la notion de "régulation"* (8409). Paris: CEPREMAP.

Lipietz, A. 1987. *Mirages and miracles: the crisis of global Fordism*. London: Verso.

Lundvall, B. 1988. Innovation as an interactive process. In *Technical change and economic theory*, G. Dosi et al. (eds), 344–69. London: Pinter.

Mackintosh Consultants 1986. *Yearbook of international electronics data 1986*. Luton: Benn Electronics Publications.

Mackintosh Consultants 1987. *Electronics data 1987 yearbook. Vol. 1, West Europe*. Luton, England: Benn Electronic Publications.

Macleod, R. M. 1986. *Technology and the human prospect*. London: Pinter.

References

Mair, A., R. Florida, M. Kenney 1988. The new geography of automobile production. *Economic Geography* **4**, 352–83.

Mansell, R. 1989. Telecommunication network-based services: regulation and market structure in transition. *Telecommunications Policy* **12**, 243–56.

Martinelli, F. 1986. *Producer services in a dependent economy: their role and potential for regional economic development*. PhD dissertation, University of California, Berkeley.

Martinelli, F. 1988. Services aux producteurs et developpement régional. In *La production des services et sa géographie*, F. Moulaert (ed.), 31–49. Special issue of *Cahiers Lillois d'Économie et de Sociologie*.

Martinelli, F. & E. Schoenberger 1990. *Oligopoly alive and well* (mimeo). Baltimore: Johns Hopkins University.

Massey, D. & R. Meegan 1978. The geography of industrial reorganization. *Progress in Planning* **10**, 155–237.

Masuyama, S. 1990. *The Pacific Rim: whither foreign direct investment?* London: Nomura Research Institute.

Maxon, J. 1988. Creating new ideas. *Management Decision* **26** (4), 40–3.

Mintzberg, H. 1988. The structuring of organizations. In *The strategy process*, J. B. Quinn et al. (eds), 52–70. Englewood Cliffs, NJ: Prentice Hall.

Morgan, K. 1989. Telecom strategies in Britain and France: the scope and limits of neo-liberalism and dirigisme. In *Strategies for new technology*, M. Sharp & P. Holmes (eds), 48–63. Oxford: Philip Allan.

Morgan, K. & A. Davies 1989. *Seeking advantage from telecommunications: regulatory innovation and corporate information networks in the UK*. Final Report to the BRIE/OECD Telecom User Group Project.

Morgan, K. & R. Mansell (in press). The coming intelligent network. Regulatory regimes, innovation and the new telecommunications paradigm. In *Regulation, innovation and spatial development*, P. Cooke (ed.). London: Sage.

Morgan, K. & D. Pitt 1988. *Coping with turbulence: corporate strategy, regulatory politics and telematics in post-divestiture America*. Working Paper 1 CICT-SPRU. Brighton: University of Sussex.

Morgan, K. & A. Sayer 1988. *Microcircuits of capital*. Cambridge: Polity.

Morgan, K. & D. Webber 1986. *Divergent patterns: political strategies for telecommunications in Britain, France and Federal Republic of Germany*. Government & Industry Working Paper 6. Brighton: University of Sussex.

Morgan, K., B. Harbor, M. Hobday, N. von Tunzelmann, W. Walker 1989. *The GEC-Siemens Bid for Plessey: the wider European issues*. Working Paper 2, CICT-SPRU. Brighton: University of Sussex.

Moulaert, F. (ed.) 1988. *La production des services et sa géographie*. Special Issue of *Cahiers Lillois d'Economie et de Sociologie*. Lille: CERIE.

Moulaert, F. 1990. *The third technological paradigm and the consultancy revolution* (working paper). Lille: CERIE.

Moulaert, F. & E. Swyngedouw 1989. A regulation approach to the geography of flexible production systems. *Environment and Planning D* 7, 327–45.

Moulaert, F. & F. Vandenbroucke 1983. Bestrijding van de werkloosheid: de bijdrage van Post-Keynesiaanse economen. In *Macroeconomics and policy*, W. Van Ryckegham (ed.), 127–54. Dordrecht: Kluwer.

Moulaert, F., Y. Chikhaoui, F. Djellal 1988. Les conseils en informatique et (télé)communications. In *La demande de services complexes des firmes multinationales et l'offre correspondence*. ERMES. Lille: Université de Lille I.

Moulaert, F., Y. Chikhaoui, F. Djellal 1991. The locational behavior of French high technology consultancy firms. *International Journal of Urban and Regional Research*.

Moulaert, F., F. Martinelli, F. Djellal 1990. *The role of information technology consultancy in the transfer of information technology to production and service organizations*. Report prepared for NOTA. The Hague: Office of Technology Assessment of the Dutch Parliament.

Moulaert, F., F. Martinelli & F. Djellal (forthcoming). The functional and spatial division of labour of information technology consultancy firms in Europe. In *Regulation, innovation and spatial development*, P. Cooke (ed.). London: Sage.

Moulaert, F., E. Swyngedouw, P. Wilson 1988. The geography of Fordist and post-Fordist accumulation and regulation. *Papers of the Regional Science Association* 64, 11–23.

Mytelka, L. (ed.) 1990. *Strategic partnerships: states, firms and international networks*. London: Pinter.

Noble, D. 1979. *America by design*. New York: Alfred A. Knopf.

Noyelle, T. 1986. *Services and the world economy: towards a new international division of labour*. Conservation of Human Resources. New York: Columbia University mimeo.

Noyelle, T. 1989. Les services et la nouvelle économie: vers une nouvelle segmentation du marche du travail. In *La production des services et sa géographie*, F. Moulaert (ed.), 3–27. Special issue of *Cahiers Lillois d'Économie et de Sociologie*.

O'Brien, P. 1989. The European computer industry – a mismatch with industry trends? Paper presented at Science Policy Research Unit, University of Sussex, July.

OECD [Organization for Economic Co-operation and Development] 1985. *Les logiciels: l'emergence d'une industrie*. Paris: OECD.

OECD 1986. Indicateurs de la science et de la technologie, no. 2, R-D, invention et competitivité. Paris: OECD.

OECD 1988a. Les accords de co-operation technique entre firmes indépendantes. *STI revue*, 4. Decembre. Paris: OECD.

OECD 1988b. *The telecommunications industry. The challenges of structural change*. Paris: OECD.

References

OECD 1988c. *New telecommunication services. Videotex development strategies*. Paris: OECD.

OECD 1988d. *New technologies in the 1990s: a socio-economic strategy*. Paris: OECD.

Ohmae, K. 1985. *The rise of Triad power*. New York: Harper & Row.

Ohmae, K. 1989. The global logic of strategic alliances. *Harvard Business Review* **69**, 143–54.

PA Consultants 1987. *Manufacturing into the late 1990s*. London: HMSO.

Pavitt, K. 1984. Sectoral patterns of technical change. Towards a taxonomy and theory. *Research Policy* **13**.

Pavitt, K. 1986. Chips and trajectories: how does the semiconductor influence the sources and directions of technical change? In *Technology & the human prospect*, R. M. Macleod, 112–34. London: Pinter.

Pelata, P. & P. Veltz 1984. Industries et térritoire: un rapport ou mutation? Le cas de l'électronique. Paris: CERTES mimeo.

Pelata, P. & P. Veltz 1985 (June). *Schémas de production et espace économique. Du Taylorisme à la production intensive en intelligence: les industries électriques et électroniques*. Paris: CERTES.

Perlmutter, H. & O. Heenan 1986. Thinking ahead – co-operate to compete globally. *Harvard Business Review* **66**, 135–56.

Perrin, J. 1987. Le phénomène Sophia–Antipolis dans son environnement régional. In *Milieux innovateurs en Europe*, P. Aydalot (ed.), 283–302. Paris: GREMI.

Piore, M. J. & C. F. Sabel 1984. *The second industrial divide*. New York: Basic Books.

Pisano, G., M. Russo, D. Teece 1988. Joint ventures and collaborative arrangements in the telecommunications equipment industry. In *International collaborative ventures in US manufacturing*, D. Mowery (ed.), 23–70. Cambridge: Ballinger.

Porter, E. P. 1980. *Competitive strategy*. New York: The Free Press.

Porter, M. E. 1985. *Competitive advantage*. New York: The Free Press.

Porter, M. E. (ed.) 1987. *Competition in global industries*. Boston: Harvard Business School Press.

Porter, M. 1989. *The competitive advantage of nations*. Boston: Harvard Business School Press.

Porter, M. & M. Fuller 1986. Competition and global strategy. In *Competition in global industries*, M. Porter (ed.), 12–32. Boston: Harvard Business School Press.

Pottier, C. & P. Y. Touati 1982. Localisation et qualification de l'emploi dans une activité à technique évolue. L'industrie micro-électronique en France. *Revue d'Économie Régionale et Urbaine*. 3.

Ribault, T. 1988. Des bases de données aux services complexes d'information. In *La demande de services complexes des firmes multinationales et l'offre correspondente*. ERMES. Lille: Université de Lille.

Roobeck, A. J. M. 1988. Telecommunications: an industry in transition. In *The structure of European industry*. (2nd edn), H. W. de Jong (ed.), 172-94. Kluwer.

Rosenberg, N. 1976. *Perspectives on technology*. Cambridge: Cambridge University Press.

Rosenberg, N. 1982. *Inside the black box*. Cambridge: Cambridge University Press.

Rosenberg, N. & L. E. Birdzell 1986. *How the West grew rich*. New York: Basic Books.

Savary, J. 1989 (September). Des stratégies multinationales aux stratégies globales des groupes en Europe. Colloque international: les groupes industriels et financiers et l'intégration européenne. Toulouse.

Schumpeter, J. 1942. *Capitalism, socialism & democracy*. London: Allen & Unwin.

Scott, A. 1988. *New industrial spaces*. London: Pion.

Scott, A. J. 1988. *Flexible production systems and regional development: the rise of new industrial spaces in North America and Western Europe*. Research paper 168. Los Angeles: Department of Geography, University of California.

Scott, A. J. & M. Storper (eds) 1986. *Production, work, territory*. London: Allen & Unwin.

Scott, A. J. & M. Storper 1987. Industries de haute technologie et développement régional: revue critique et reformulation theorique. *Revue Internationale de Sciences Sociales* **112** (May).

Stigler, G. 1951. The division of labor is limited by the extent of the market. *The Journal of Political Economy* **59**, 185-93.

Swyngedouw, E. 1987a. Social innovation, organization of the production processes and spatial development. *Revue d'Economie Urbaine et Régionale* **3**, 487-510.

Swyngedouw, E. 1987b. *Telecommunications and regional development: an assessment of the role of telecommunication industries and its implications for regional development patterns*. Report prepared for the EEC, DG IV, Research Programme on European Integration, Brussels.

Swyngedouw, E. 1988. The geography of high technology production and the defense/technology nexus. *l'Espace Géographique* **12**, 153-71.

Swyngedouw, E. & S. Anderson 1987a. Les dynamiques spatiales des industries de haute technologie en France. *Revue d'Economie Urbaine et Régionale* **2**, 321-49.

Swyngedouw, E. & S. Anderson 1987b. Le schéma spatial de la production de haute technologie en France. *Revue d'Economie Urbaine et Régionale* **2**, 409-37.

References

Teece, D. 1986. Profiting from technological innovation: implications for integration, collaboration, licensing and public policy. *Research Policy* **15**, 285–305.

Thackery, J. 1987. DEC's sweet revenge. *Management Today* (December), 64–70.

Thrift, N. J. & A. Leyshon 1988. "The gambling propensity": banks, developing country debt experiences and the new international financial system. *Geoforum* **19** (1), 55–69.

Ungerer, H. & N. Costello 1988. *Telecommunications in Europe*. Brussels: CEC.

Valeyre, A. 1982. Dynamique régionale de l'emploi et division spatiale du travail. *Revue d'Économie Urbaine et Régionale* **3**, pp?.

Vliet, A. van de 1987. Finland's wide awake giant. *Management Today* (July) 62–7 & 108–9.

Von Hippel, E. 1987. Co-operation between rivals: informal know-how trading. *Research Policy* **16**, 291–302.

Weinstein, O. 1983. Cycles longs, mutations et crise. *Issues* **16**.

Weinstein, O. 1988. Production et circulation des connaissances scientifiques et technologiques: la recherche-développment comme activité économique spécifique. In *La production des services et sa geographie*, F. Moulaert (ed.), 59–80. Special issue of *Cahiers Lillois d'Économie et de Sociologie*.

Williamson, O. 1985. *The economic institutions of capitalism*. New York: The Free Press.

Wiseman, C. 1985. *Strategy and computers: information systems as competitive weapons*. New York: Dow-Jones, Irwin.

Womack, J., D. Jones, I. Roos 1990. *The machine that changed the world*. London: Macmillan.

INDEX

Abernathy, W. 22, 32
Adler, P. 24
Aglietta, M. 39
Alcaltel 10, 73, 75, 152–73, 175, 198
Amin, A. 59
Andersen, P. 100, 104
AT&T 10, 17, 69–78, 124, 128–33, 163, 194
alliances 72–8, 143–7
 strategic 72–8, 143–7
 value-added partnering 78
Arnold, E. 128

Barreau, J. 151
Beckouche, P. 102
Boyer, R. 2, 22, 28, 39, 40–7
Bretton Woods 46
British Telecom 151–73, 198 210
Brittany 15, 100–128
Brooks, H. 61
Bull 7, 17, 130, 140–68, 200

C&C 1–5, 9, 15–18, 21, 39, 47–57, 61, 71–8, 78–128, 178–99, 200–214
 computing 1–5, 8–9, 15–18, 61–78, 78–128
 hierarchical 16
 decentralised 16–17
 communications 1–5, 9, 15–18, 61–78, 78–128, 129–51, 152–77
Cohendet, P. 33–4
Cooke, P. 4, 22, 72
Coriat, B. 22, 24–5, 28, 33–4
Costello, N. 151–5
corporations 3, 208–11

hollow 47
 horizontal integration 54
 functional integration 54
collaboration 76–8
 strategy 147–77
Crandall, R. 132
customization 26

DEC 11, 76, 132–51
deregulation 69–78
dirigisme 7
Dosi, G. 32, 42, 48
Dunford, M. 213
Durand, P. 96

economies of scope 26–8
Enderwick 64
England 13–15, 84–97, 120–28
Ericsson 164–77, 202
Ernst, D. 76
European Community 18, 59, 74–5, 205
 ESPRIT 74–5, 147, 205

Flamm, K. 130–32
Fordism 6, 9, 14, 19–38, 44–60, 200–209
 post-Fordism 9, 14, 213
 Sloan-Fordism 25
 Sloanism, 25, 28
 Taylorism 32, 50, 179
 flexibility 5, 22–38
 static 5, 33
 dynamic 5, 26, 33–8, 213
France 78–128
France Telecom 17, 70, 155–68, 202
Freeman, C. 21, 48

Frobel, F. 66
Fuller, M. 76
Fujitsu 10, 76, 130, 172, 200

Gadrey, J. 58
GEC 70, 73, 90, 167–77,
 200–203
Gerwin, J. 24
Goddard, J. 58
Guile, R. 61
globalization 2–3, 10–12, 61–78,
 81–128, 163–4, 186–99
 competition 131–51
 global localization 75–8,
 200–214
 global solutions 53, 82, 139
Granovetter, M. 4
Grey, S. 67–8
Guy, K. 132

Hacklisch, C. 76
Hamel, G. 78
Hamill, J. 67
Hagedoorn, J. 76–8
Hakansson, H. 3
Harrigan, K. 76
Hart, H. 34
Heenan, R. 76
Hewlett-Packard 11, 132–51
Hills, J. 7

IBM 3, 8, 10–11, 76, 131–51, 198
ICL 8, 70–71, 76, 130–51, 172, 200,
 204–206
ISDN 153–7
ITT 7, 202
innovation 3, 31
information 5, 22–38, 56
 logistics 56
 knowledge-intensive production 5,
 22–6, 29–33, 35–8, 289
 science 30, 32–8

Janssens, F. 56

joint venture 56, 72–8,
 147–51
Jowett, P. 198
"just in time" 55, 57

Keynes(ianism), J. M. 7, 9, 45
Klein, B. 5, 33–4, 35
Kristensen, P. 22, 37

local assimilation 75
learning-by-using 32
Leborgne, D. 2, 39–40
Leyshon, A. 65–6
liberalization 71, 162
Lipietz, A. 2, 39–40
Llerena, P. 27–34
Lundvall, J. 4

Mair, A. 66
Mansell, R. 3
Martinelli, F. 54, 57–8, 197
Masuyama, S. 75–8
Matra 71, 167–77
McDermott, M. 67–8
Mercury 70–74, 158–77
Mintzberg, H. 188
Morgan, K. 4, 68, 127, 156–70
Motorola 76, 127, 150, 167
Moulaert, F. 2, 39, 57,
 179–80, 183–6, 194–5, 213
Mouline, J. 155
Mytelka, L. 4, 74–6

NEC 10
Noble 30
Nord–Pas-de-Calais 104–28

Scotland 13–15, 84–97, 120–28, 213
Scott, A. 2, 12
Siemens 10, 73–6, 138, 168–77,
 206–210
software 16, 80–128, 131,
 140–44, 182–203
Solow, R. 208

spatial change 12, 58
Stigler, G. 22, 29
Storper, M. 2
Sun Microsystems 131–51
Swyngedouw, E. 40, 55, 58, 104, 107
systems 16, 178–99
  engineering 16, 178–99
  integration 16, 80, 178–99

technology 2–5
telecommunications 7–8, 69–78, 152–77
Teece, D. 4, 25, 32, 76–8
Thrift, N. 65–6
Triad Power 10, 61

Ungerer, H. 153–5
user power 3, 205–6
Utterback, J. 32

Valeyre, A. 128
Vandenbroucke, F. 45
Veltz, P. 31, 104, 127
Von Hippel, E. 76

Wales 13–15, 91–7, 120–28
Weinstein, O. 20, 32
Williamson, O. 62
Wilson, P. 39
Wiseman, C. 62
Womack, J. 212

Zysman, J. 33–4